UNRULY!

A weird and wild history of weeds in America

Unruly

The Wild and Weird History of Weeds In America

Olivia Wylie

Leafing Out Books 2021

ISBN: 978-1-7343271-5-1

For Gram.

You gave me my first book on plants, and you helped me tape the pages back in when I cracked the poor thing's spine.

Today I write this book on plants for you, with all my love.

Introduction

"Oh, how did *you* get in here?"

It's a phrase grumbled by generations of gardeners, directed at generations of weeds (among some other choice words of the $*%$!!! variety).

"How did you get in here, you little monster?" was the phrase I used to hear my grandma mutter at some impertinent goosefoot or ruffian thistle that dared to put its head up in her nice herb plot. Thinking back, I wonder whether my grandma ever reflected on her grandmother, her great-grandma and the many women who'd been throwing those self-same plants out of their cultivated areas across several centuries and two continents.

The goosefoot, in particular, has gotten in the habit of riding humanity's coat-tails, and it can be found in every large human settlement outside the polar circles. After all, that's what makes a plant a weed: being irrepressible. Being prolific. Being innovative. And being in the business of making use of humans. This seems to annoy a great many humans, who have somehow gotten the idea that the relationship should be the other way round.

I am a garden designer and horticulturist by profession, and I've done my fair share of cursing at the weeds featured in this book. But I can't resist a grudging smile at bindweed lending grace to the poor broken ground of a construction site, or the exuberant yellow of a swathe of dandelions who have escaped beheading by the lawnmower.

Weeds are aggressive, yes. They are a definite threat: the U.S. Cooperative Extension System reports that weeds cause more yield loss and add more to farmer's production costs than insect pests, crop diseases, root-feeding nematodes or warm-blooded pests.

Weeds are a psychological threat as well: they threaten our sense of order with their constant reminder that nature is not exactly under our control. Richard Mabey said it best in his book, *Weeds: How Vagabond Plants Gatecrashed Civilisation and Changed the Way We Think About Nature*. "The wild gatecrashes our civilised domains, and the domesticated escapes and runs riot. Weeds vividly demonstrate that natural life—and the course of evolution itself—refuse to be constrained by our cultural concepts. In doing so they make us look closely at the very idea of a divided creation."

If we pay attention, the weeds also teach us lessons. Some lessons are practical: for instance, weeds can tell you what your soil has and what it needs. These green guideposts work better than the soil tests from the corner store in telling you that you happen to be standing on:

Nitrogen-rich soil:

Annual nettle (*Urtica urens*)
Chickweed (*Stellaria media*)
Common orache (*Atriplex patula*)
Cow parsley (*Anthriscus sylvestris*)
Elderberry (*Sambucus nigra*)
Fathen (*Chenopodium album*)
Goosegrass (*Eleusine indica*)
Hogweed (*Heracleum sphondylium*)
Mugwort (*Artemisia vulgaris*)
Shepherd's purse (*Capseila bursa-pastoris*)
Stinging nettle (*Urtica dioica*)

Nitrogen poor soil:

Corn chamomile (*Anthemis arvensis*)
Sheep's fescue (*Festuca ovina*)
Sweet vernal grass (*Anthoxanthum odoratum*)

Alkaline soil:

Betony (*Stachys officinalis*)
Blue moor grass (*Sesleria varia*)
Forking larkspur (*Consolida regalis*)
Meadow clary (*Salvia pratensis*)
Pheasant's eye (*Adonis aestivalis*)
Salad burnet (*Sanguisorba minor*)
Sanicle (*Sanicula europaea*)

Acid soil:

Bracken (*Pteridium aquilinum*)
Corn chamomile (*Anthemis arvensis*)
Creeping soft grass (*Holcus mollis*)
Heather (*Calluna vulgaris*)
Sheep's sorrel (*Rumex acetosella*)
Wavy hair grass (*Deschampsia flexuosa*)

Compacted soil:

Dandelion (*Taraxacum officinale*)
Field horsetail (*Equisetum arvense*)
Greater plantain (*Plantago major*)
Silverweed (*Argentina anserina*)

Wet soil:

Coltsfoot (*Tussilago farfara*)
Compact rush (*Juncus conglomeratus*)
Corn mint (*Mentha arvensis*)
Creeping buttercup (*Ranunculus repens*)
Purple moor grass (*Molinia caerulea*)
Soft rush (*Juncus effusus*)
Wood club rush (*Scirpus sylvaticus*)

Dry soil

Broad-leaved thyme (*Thymus pulegioides*)
Bugloss (*Anchusa arvensis*)
Golden marguerite (*Anthemis tinctoria*)
Whitlow grass (*Draba verna*)

Not only do these plants tell us the condition of the soil at our feet, they tell us the history. According to Emerson, a weed is a "plant whose virtues have not yet been discovered." Even the word has changed its values over time; the word *weed* is from Old English, dates back to the 7th century, and is derived from *weod*. In the beginning the word referred to all plants—including grasses—that grew together in a meadow. By the Middle Ages it had taken on its current meaning. Around 1600, the word was borrowed for slang to describe tobacco, as it was in the 1920s to describe marijuana. Even our language can't decide what a 'weed' is.

I'd like to amend Emerson's statement: a weed is a plant whose virtues have *been forgotten* or have not been discovered. We know what our ancestors employed as salad greens or medicine by what they brought to this country and set loose, abandoned, or allowed to escape their gardens. The use of the tumbleweed to visually define the landscape of the American West tells us that our cultural image of the area comes from a period after the 1860s, when its seeds hitched a ride in bags of flax brought by Russian immigrants to South Dakota. The cheery dandelion that once kept scurvy at bay for English settlers is cussed by the descendents of the people that the plant nourished. They say that you can trace the Oregon Trail by following the bright yellow flowers of Harison's Yellow Rose across the prairie. Wild thickets of tree of heaven plants, Siberian elm, and Japanese knotweed rub shoulders in abandoned New York city lots with plantain, horseweed, pigweed, Oriental bittersweet and Amur peppervine. The plants attest in their tangle to the number of nationalities whose goods—and weeds—pass through the city's ports. You can see the effects of human hubris in Bradford pear and the danger of ecological experimentation in kudzu, the dread 'Weed that Ate the South'.

In this book, I ask my grandma's question—"how did you get here?"—in all seriousness. I'll explore the histories of American invasive weeds in all their cussed, crafty, and cantankerous glory. I'll explore the histories of invasive weeds that are either among the most ubiquitous, or the most dangerous species in the United States. How did they get here? We brought them; by intent, by accident, by delusion or by design. When we listen to their story, we learn not just what *they* are, but what *we* are: hopeful, driven, careless, wandering. Dreaming, greedy, innovative. Selfish, sentimental, shortsighted, and full of so much potential.

Table of Contents

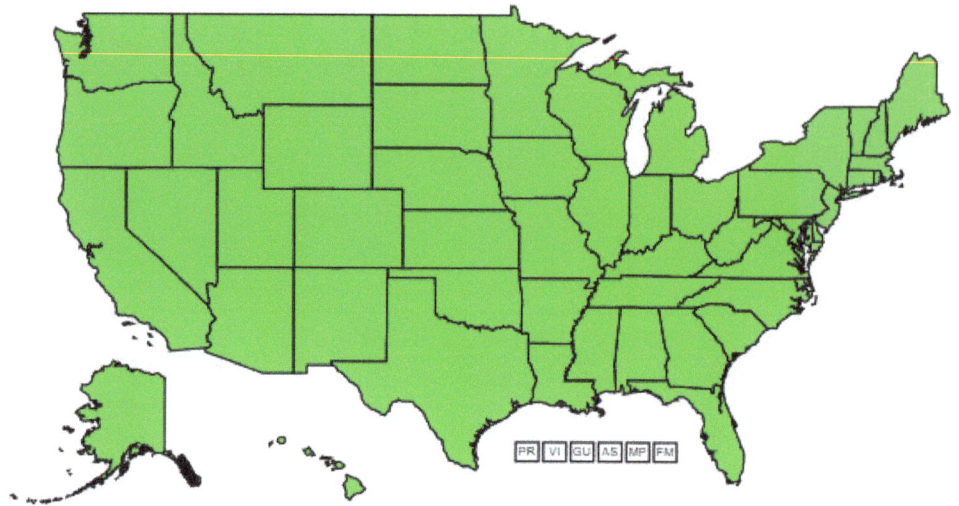

Legend
- List only
- Law only
- Both List and Law
- Included on other source
- Not Listed

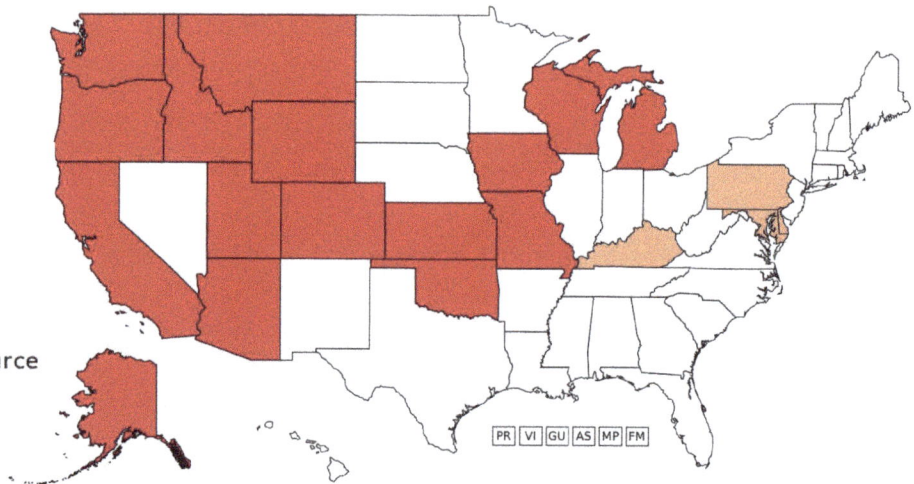

A Note on The Maps Shown in The Book

The two maps you will see with each entry are courtesy of the EDDMapS program. EDDMapS is a web-based mapping system for documenting invasive species and pest distribution. It is fast, easy to use, and doesn't require Geographic Information Systems experience. Launched in 2005 by the Center for Invasive Species and Ecosystem Health at the University of Georgia, it was originally designed as a tool for state Exotic Pest Plant Councils to develop more complete distribution data of invasive species. Since then, the program has expanded to include all invasive species in the U.S. and Canada, as well as to document certain native pest species.

EDDMapS' goal is to maximize the effectiveness and accessibility of the immense numbers of invasive species and pest observations recorded each year. As of February 2021, EDDMapS has over 5.1 million records.

EDDMapS aggregates data from other databases and organizations as well as volunteer observations to create a national network of invasive species and pest distribution data that is shared with educators, land managers, conservation biologists, and beyond. This data will become the foundation for a better understanding of invasive species and pest distribution around the world.

The State Distribution maps: Shown in shades of green, these maps are based on current site and county level reports made by experts, herbaria, and literature.

The State Impact maps: Shown in shades of red, these maps identify those states that have the species in question on their invasive species list or laws.

To learn more, please visit https://www.eddmaps.org

Bindweed

Convolvulus arvensis

Common Names: This plant has annoyed enough people to have 84 names in 29 languages. In America, its most common names are bearbine, bethbine, cornbine, field bindweed, creeping jenny, possession vine, devil's guts, smallflowered morning glory, hedge bells, and corn lily.

Family: Convolvulaceae

Genus: *Convolvulus*

Looks Like: a perennial, broad-leaved deciduous plant no more than 8 inches high. Lanceolate, alternate leaves are arranged on vining tendrils that arise from a large central root that acts as a nutrient reservoir. These vines wind their way around any available support, and can be up to 8 feet long. They form dense, webbed mats.

Leaves are up to three inches long by an inch wide, resembling arrowheads. The half-inch to two-inch wide flowers resemble small trumpets; often cream-colored, sometimes tinged or striped in pink. These flowers age into a rounded fruit that contains from two to four seeds, resembling a tiny Victorian street lamp. The seeds are viable in the soil for several decades.

Comes From: Southern Europe, around the Mediterranean. It's harder to get more specific than this with such a prolifically spreading plant. It's been travelling with humanity long enough to be mentioned in the writings of Dioscorides, a Greek medical herbalist. Not in a nice way, either.

Likes To Live: just about everywhere outside the polar ice caps. It has reportedly been found in the Himalayas, at 10,000 feet. It can live in any type of soil, though it does prefer sun, and will instantly colonize cleared earth such as construction sites or agricultural fields.

States Present: All 50 states, Puerto Rico and the Virgin Islands.

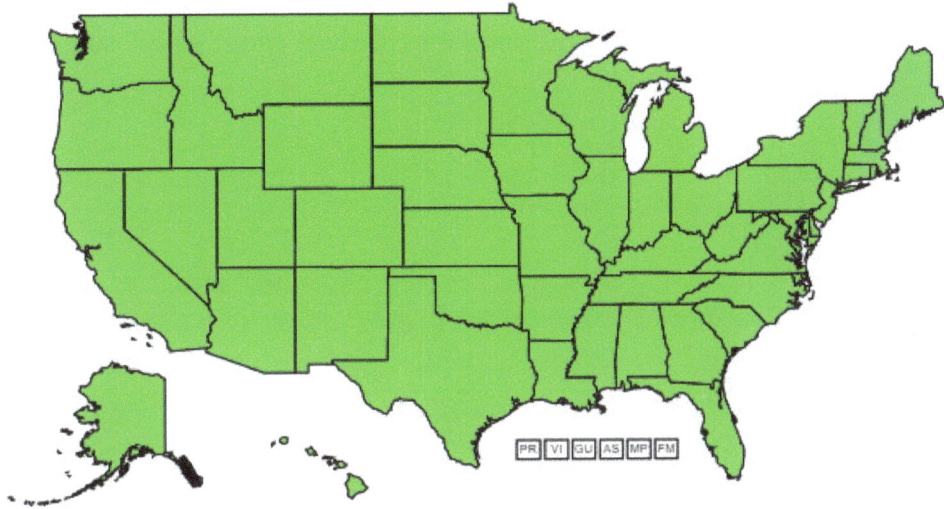

States Impacted: States with dryland crop or vineyard production.

Legend
- List only
- Law only
- Both List and Law
- Included on other source
- Not Listed

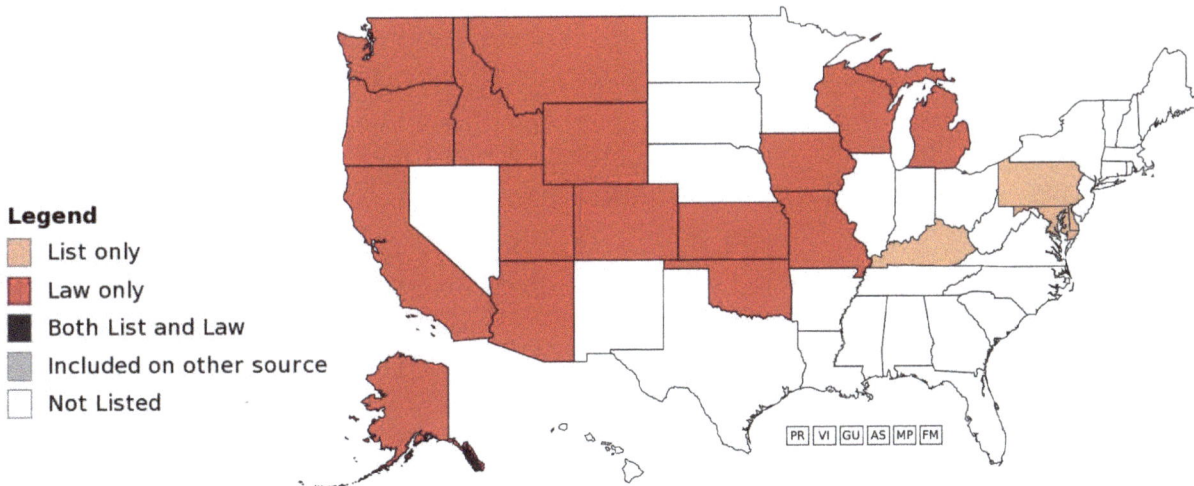

5

Succeeds By: the triple threat of deep, extensive horizontal root systems, prolific seed production and root buds that develop into rhizomes and shoots when exposed to light. Field bindweed flowers from June to September through most of its range, and a single plant can produce 550 seeds in a season. One of the plant's less common nicknames is Lazarus Weed, because each fragment of root left when the plant is yanked will grow into a new plant. One plant will grow a root system that may be 10 feet deep and contain two or three years' worth of nutrition held in reserve. This makes eradicating it a challenge.

Causes Trouble By: growing over and smothering other plants as it climbs them to get sunlight. Outcompeting other plants for nutrients. It cuts down on biodiversity by outcompeting native plants that provide food for wildlife. Bindweed can harbor the viruses that cause potato virus X disease, tomato spotted wilt, and vaccinium false bottom. The vines also make a host for powdery mildew. The plant is the bane of agricultural fields, where its threadlike vines can tangle in and clog farm machinery. It cuts down on yields by scrambling up (or over) crops and binding them together, making the plants up to 50% less productive while driving the farmer crazy.

Carl Linneaus gave this plant the official name of *Convolvulus arvensis* in his seminal book, *Species Plantarum*, published in 1753. You couldn't have come up with a better name for the plant. *Convolvere* is a Latin verb meaning 'to entwine', and *arvens* is Latin for 'of the field'. 'The entwining field plant' pretty much sums bindweed up. There's a lot more you could say about it, such as 'gets everywhere', 'I just pulled you!' and 'stubborn little monster.'

Based on the somewhat spotty historical record, bindweed has been with us as long as we've been working the land. A native of the Mediterranean, it was noted by Dioscorides in the first century AD He recommended it, drunk as a tea, to cure spleen problems, weariness, and hiccups. However, he warned, it had the inconvenient side effects of causing one to urinate blood after the 6th day's dose and making one permanently sterile after the 37th day. Maybe *not* one to try at home.

It was occasionally used as a laxative over time, and was sometimes called 'poor man's Jalap'. But it's not great for the purpose. It does have one decent use: if you like to dye your own wool, bindweed's fleshy white roots do give a nice, solid yellow color. But other plants do it better, without strangling their neighbors.

So why did this little Mediterranean monster become something worth importing to the New World? Well, it was only intentionally imported now and then. The rest of the time, it imported itself. Bindweed came here using a number of routes. One of its tools was, believe it or not, beauty. Like a naughty toddler, we can't help but fall for the charm of this little white flower, even when we despair of its behavior. As late as 1807, it was sold by mail-order seed companies for use in hanging baskets and as—rather ironically—a ground cover. In fact, field bindweed is rumored to have been established in the Pacific Northwest by an Oregon settler who used it as a cover crop in his orchard. I imagine he wasn't too popular with his neighbors.

But bindweed didn't arrive by the choice of gardeners alone. No, it made its own way over the sea. Stowing away in cereal crops, its seed was broadcast along with wheat and rye. Carelessly packed into the fodder of animals being shipped to the New World, it was—ahem—planted by the digestive systems of livestock. The little bindweed seeds are plenty tough enough to traverse a cow's system and germinate on the other end. First documented in the United States in Virginia circa 1739, it was noted as a pest in Pennsylvania circa 1812, and in Maine around 1824. By 1919, it was declared as 'the Worst Weed of California'. Between 1900 and 1970, a flood of articles and periodicals were written on the eradication of field bindweed.

So many words. So little success. In spite of its listing as a noxious invasive and mandated extermination by law in two thirds of the United States, it still sprawls happily anywhere it can find bare dirt and sun. Its shining white blooms give grace back to the broken ground of construction sites and the scars of disturbed land. Like a cheeky kid, bindweed smiles up at us and scrambles away, irrepressible.

If You Love It: Leaving bindweed to climb fences or colonize bare areas isn't a terrible idea. Just check your city regulations and make sure it isn't going to get you a ticket. Check the status via USDA-NRCS. It's usually down as a lower-level noxious weed or list C weed, which means local governments often require it to be contained, eradicated or suppressed. But it is a close relative of morning glories (who can be a little weedy in their own right) and has a generally similar manner. Like its more decorative relations, it is a food source for bees and small butterflies.

There aren't particularly reliable recorded uses for the plant. Bindweed contains several alkaloids, including pseudotropine, and lesser amounts of tropine, tropinone, and meso-cuscohygrine. This makes it pretty unhealthy as a food source. Boiling the leaves will remove the worst of the alkaloids, and bindweed can be treated as an early spring green... as long as you're careful where you harvest. Stick to your own backyard. This plant is religiously sprayed in many places, and ingesting herbicides is a bad idea.

If You Loathe It: Should bindweed be driving you to distraction, let's talk control methods.

For the Home: The absolute best control in the home garden is thick mulching, spread over weed fabric in the cases of really bad infestations. Pin weed fabric down with fabric pins, and spread the mulch over it to a thickness of 3 inches (about the second joint of your index finger.) Whatever works its way through that should be pulled by hand. Expect some bindweed to pop out anywhere it finds an edge or a hole. This is the way it tries to survive. Be vigilant!

In areas like brick walkways, I personally use a small flame wand sold in the tools section of your local hardware store. Use it with a little common sense: don't use the flame wand in areas with lots of dry materials that could ignite. But for cracks in concrete, brick walkways and gravel paths, it's just the thing. This tool will not kill the roots of bindweed, but since not much will, it's a lot better than getting on your knees with a weed-pulling tool and trying to work a blade between bricks!

A word of warning: a lot of people will tell you to spray bindweed with vinegar, dish soap, and epsom salts mixed in hot water. Don't. It's bad for your soil, bad for your sprayer, and it won't do much to bindweed. If you don't want to use chemicals, go with iron sprays, mulching, and the fire wand. The most common biological control agent for the control of field bindweed is *Aceria malherbae*, the field bindweed mite. The insects seem to prefer dry, arid soils. Mites establish well on roadsides and in parking lots, just like the bindweed. But their effects are wildly variable, and research on their effectiveness is very mixed.

The best tool you have against field bindweed is vigilance. Keep after it. Wear it down. Over two or three years of constant work, you will exhaust the plant's reserves and it will finally be gone from your property. Smile at it when it scrambles across vacant lots and empty construction sites, but don't give it any space in your garden. Give bindweed an inch, and it'll take the yard.

The Chemical Option: If you have a truly obnoxious infestation, you may need to pull out the big guns. 2,4-D and dicamba are both effective at killing the plant, as is glyphosate (the active ingredient in Roundup) to a lesser degree. If you choose dicamba, please read the label carefully! I advise that you spray early in the day. It can turn into a gas that drifts and damages other plants when the temperatures are at or above the 75F range. Glyphosate is probably going the way of DDT as we come to understand it better, and I wouldn't recommend it.

Black Medic

Medicago lupulina

Common Names: black medick, nonesuch, bur clover or hop clover, black clover, black hay, blackweed, poor man's alfalfa and English trefoil.

Family: Fabaceae

Genus: *Medicago*

Looks Like: a prostrate broadleaf annual or tender perennial. The plant grows close to the ground, spreading up to 2 feet in a dense mat. The plant's dark green leaves are similar to clover leaves, with three oval leaflets. Each 3/4" long leaflet has a spur on the tip, toothed margins, and parallel veins. The center leaflet protrudes just a little. This and the toothed margins help to distinguish black medic from other clovers. The leaves are laid out alternately along the stems. The small, bright yellow flowers are produced on the leaf axils. Each inflorescence is a compact, rounded cluster of 10 to 50 tiny flowers. Flowers show up throughout the growing season, although individual plants stop blooming once seed is set. The fruits that form after pollination look like small kidneys, arranged in clusters. The coiled seed pods turn black when ripe. Each seed pod contains a single golden or brown seed.

Comes From: the Mediterranean basin, Southern Ethiopia, Kenya and Tanzania.

Likes To Live: in dry, sunny areas like roadsides and railroads. It can be a nuisance in gardens, lawns and fields as well.

States Present: All 50, even Alaska!

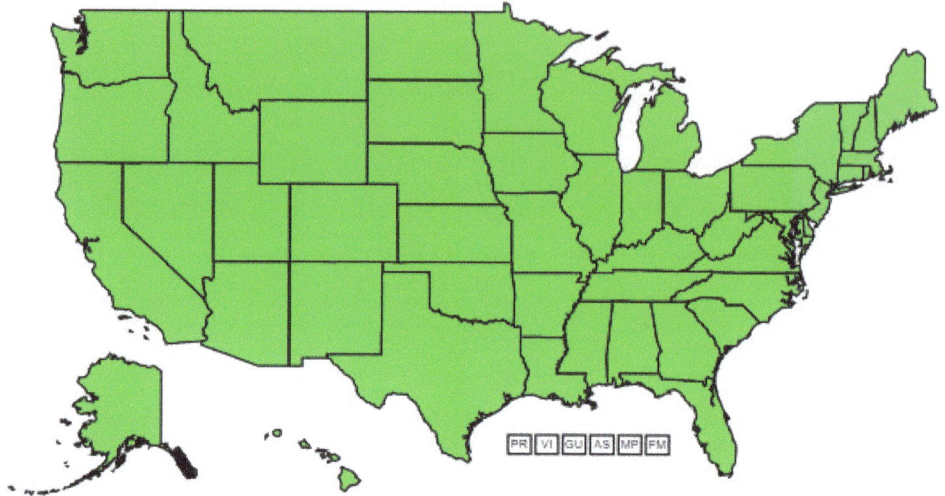

States Impacted: Black medic can be a weed in turfgrass and in ornamental plant beds. It thrives in nutrient-poor soil and during dry conditions. While it can be a nuisance, it is rarely economically problematic and can do some good: as a legume, it fixes nitrogen and can enrich the soil.

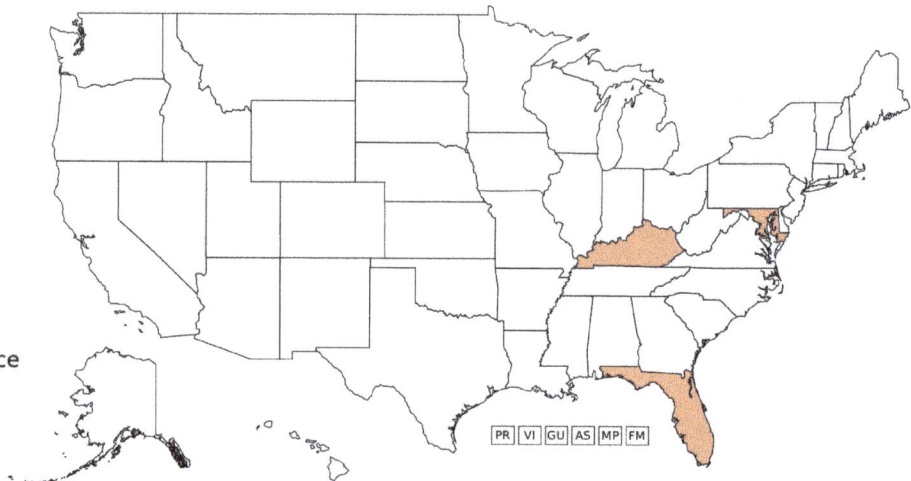

Legend
- List only
- Law only
- Both List and Law
- Included on other source
- Not Listed

12

Succeeds By: producing viable seed under the height of most mower blades. These seeds can persist in the soil for years.

Causes Trouble By: outcompeting a *lot* of other plants when the soil gets hot, dry and runs low on nutrients.

Medicago lupulina. They aren't long words, but a very long history is wrapped up in them. Let's start with *Medicago*. *Medicago* (med-ik-KAY-goh) is Hellenized Latin. Why this happened, sources differ. Some say the Greeks imported alfalfa and its little cousin from what they called Medea, the place we call Iran today. Some say the Medes brought it with them when they invaded Greece. However it happened, the Greeks called it "Medea Grass" which in Greek is μηδική (pronounced mee-thee-KEE). The Romans called it *Medika*. Through linguistic drift and some Anglicized misunderstandings on how to write Latin, the word ended up as '*medicago*'.

Lupulina is Latin for Little Wolf, for a very odd reason. The black medic blossom resembles hops. In Latin, Hops is *Humulus lupulus,* the Wolf of the Earth. Since Black Medic resembled the plant, it was given a similar name. Funny what threads history weaves together, isn't it?

I debated including black medic among the weeds at all, since it is such a valuable little plant in rangeland, animal forage, and orchard management. In fact, in some circles there are strong arguments made for rehabilitating its reputation, and actually increasing use of it in cover-crop seed mixes. It was probably introduced to North America in the late 1700s or early 1800s, in general pasture seed mixes designed to create nutritious and sustainable forage for pastures. A fast grower and a nitrogen fixer, the little plant supports other forbs, and it grows back quickly when, say, trampled by cows, which makes it a wonderful soil protector.

For that matter, it's a great food for humans, too. According to the lovely book *Medicinal Plants of China* by Dr. James Duke, three ounces of black medic forage has about 23 grams of protein and 3 grams of fiber. A word of caution: If you do decide to eat black medic, cook it as a pot herb rather than eating it raw. It contains L-canavanine, especially in the seeds and sprouts. L-canavanine is an amino acid which can cause abnormal blood cell counts, spleen enlargements, and can be a flare-up trigger for lupus sufferers. It's a good idea to avoid it if you're pregnant or on blood-thinners.

There isn't a lot of folklore on black medic, though it did charm Cicely Mary Barker enough to be included in her famous flower fairy depictions. It has one distinction, though. It's said that when Saint Patrick stood and gave his speech on the Trinity so long ago, he plucked a small shamrock to illustrate the concept. Some say it was black medic he held. If that tale is true, then why is the four leafed clover so revered? Well, that's an interesting little story. Aside from the fact that four-leaved forms of the plant are rare and we humans are magpies for rarity, there is an interesting bit of lore: four leaves stand for faith, hope, love and luck. A wonderful find, indeed!

If You Love It: Black medic actually has a lot of great qualities! It has long been used in orchards and vineyards for soil coverage, and helps preserve the organic content of the soil thanks to its ability to fix nitrogen. It can be used for the remediation of soil that is out of nutrients, and can help clean up high levels of heavy metals such as lead. It's used to preserve soil moisture and prevent weed development in European farms.

Black medic is mainly sown in pastures. It mixes well with grasses or other clovers to make good quality forage for animals. If you'd like to plant it, spread the seeds at a rate of 9 pounds per acre.

If You Loathe It: You're not the only one. It's a real nuisance in lawns, and if your job is keeping the grass looking good it's a pain to fight with. So, here are some tips.

For the Home: The best thing for control is good soil care. Black medic outcompetes grasses and ornamentals when the other plants are struggling in difficult conditions. Amend the soil with organic material and good quality mulch to prevent the loss of moisture, and you're already well on your way to giving other plants a chance to win the competition. Hand removal is also an option in ornamental beds, particularly when the soil is moist. Since it is so low-growing, there is one more option in turf: mow it a little high. The turf will shade the black medic, and it can do its good nitrogen fixation work out of sight.

The Chemical Option: If you are going to use chemicals—and I'd only recommend this in, say, a one-acre yard that's been allowed to go to seed (pardon the pun) for a few years—stick to the pre-emergents dithiopyr (Dimension), and oxyfluorfen and oryzalin (Rout). For post-emergents, I wouldn't go for anything any stronger than 2,4-D.

Note: Chemical control should only be used as a last resort, as organic approaches are safer and much more environmentally friendly. Of course, you could just ignore it, keep the turf longer and let black medic do its nitrogen fixing out of sight, couldn't you? Just a thought.

Bradford Pear

Pyrus calleryana 'Bradford'

Common Names: Callery pear, bird pear, tuna on a trunk (read below and you'll find out why. Hold your nose!)

Family: Rosaceae
Genus: *Pyrus*

Looks Like: a fine tree in domestic settings, straight-limbed and deciduous, growing 16 to 26 ft tall with good form and a conical canopy. In its feral state, it tends to form prickly thickets. The leaves are oval, 4 to 8 cm (1-1/2 to 3 in) long. The bark is smooth and charcoal-grey. The white, five-petaled flowers appear early in spring, about 3/4 to 1 inch in diameter. The domestic plants are thornless; the wild escapees arm themselves with thorns as long as your pinky.

Comes From: China and Taiwan.

Likes To Live: in well-drained loam with consistent moisture, in full sun. Tolerates some drought once established. Adaptable to a wide range of soil conditions, including heavy clays. The tree is surprisingly adaptable, and extremely tolerant of urban conditions. It thrives in poor soil, wet or dry, acidic or alkaline.

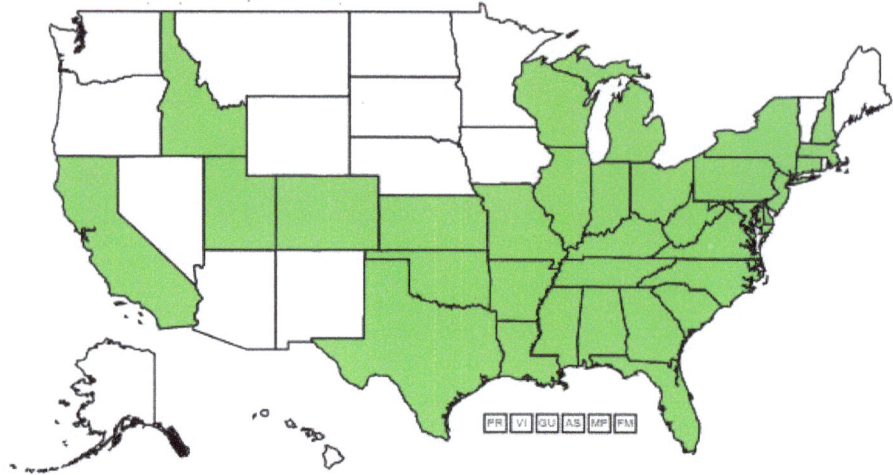

States Present: 152 counties in 25 states.

States Impacted: Illinois, Tennessee, Alabama, Georgia, South Carolina, Virginia, West Virginia, Pennsylvania, Delaware, New York, Massachusetts, Connecticut, Indiana, Arkansas. Arkansas pays a bounty to anyone who cuts them down!

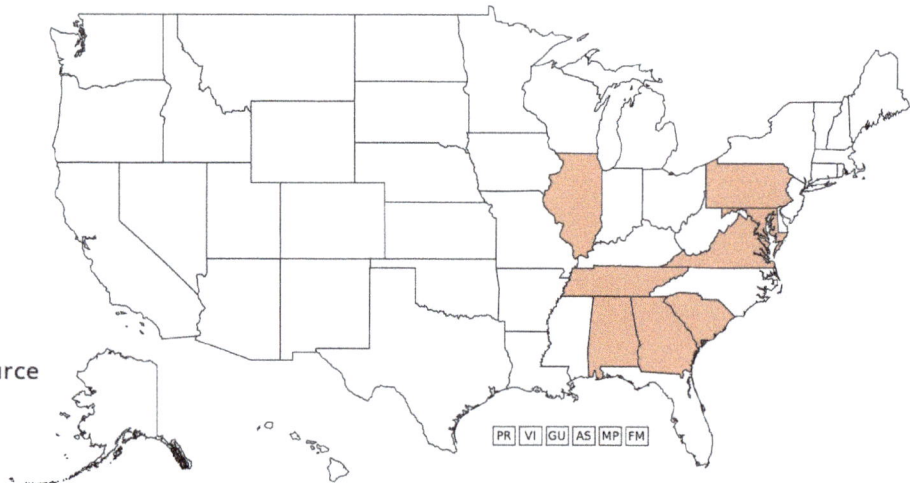

Legend
- List only
- Law only
- Both List and Law
- Included on other source
- Not Listed

Succeeds By: being sneaky and tricking overly optimistic humans. The originally cultivated trees were unable to self-pollinate and produced sterile fruits. They were widely planted throughout the United States as an ornamental. But when the trees hybridized with other types via windborne pollen, they produced fertile fruit. To make matters worse, fertile pear varieties are commonly used as the rootstock when grafting. If the grafted crown is damaged, the fertile rootstock can dominate, producing fertile fruit. Which means the tree seeded into the wild. Its originally useful traits then made it a pest: ease of adaptation and resistance to all pests and disease.

Causes Trouble By: squeezing out native flora and reducing biodiversity, quickly turning meadows into thickets of thorny trees. It invades riparian areas, farm fields, and anywhere where birds can drop its seeds or its suckers can reach.

If I had my way, another common name for Bradford pear would be 'hubris tree.' In 1908 it seemed like a retail plant breeder's dream made real, too good to be true. But you know the old adage: if something seems too good to be true, it probably is.

The Bradford pear has quite a story behind it. The introduction of the Bradford pear in America began as a way to combat a problem: fire blight. This nasty infection caused by the bacterium *Erwinia amylovora* hits members of the rosacea family particularly hard. Sadly, this included pear and apple trees. Around the end of the 19th century, Northern California and Southern Oregon had become busy centers of pear production in the United States. Pears were a high-value fruit. How high-value? Consider this. In 1916, the income from pear harvest for a single Oregon county was $10 million. That's $28 million in today's money. Pears were big business. And that business was in danger.

Enter the plant scientist Frank Reimer, who thought he might just have the answer. He knew of a Chinese pear that was highly resistant to fireblight, and asked the energetic young botanist David Fairchild to acquire a number of samples for planting trials; if the new tree was a success, it could be used as a rootstock, and grafting could save the pear industry.

The work of travelling China in search of this pear was in turn handed by Fairchild to Frank Meyer. Meyer, known both for his skills in botanic discovery and for his eccentric, mercurial nature, set about scouring southern China. There is a powerful picture of him in archives today, dressed for the harsh terrain and gazing into the future with burlap-swaddled saplings laid about his feet like sleeping children. By 1918, Meyer would send home hundreds of seed and sapling specimens of what would become Callery pear, along with many other species. But he was a troubled man, and his death by suicide prevented him from seeing what became of his work.

At first, it seemed that something bright had grown out of a murky past. First, the new pear served as an excellent rootstock. Then the horticulturist John L. Creech found a particularly stunning specimen among the possible source trees from Nanjing province. He was struck by its lovely form and its perfect disease resistance. Even better, the tree did not have the thorns of other Callery pears. It was beautiful. It was tough. It was clean. It was perfect. Creech was in love. He selected the tree as the source for hundreds of clones. In selecting this individual to mass-produce, Creech named the cultivar after the former head of the USDA Plant Introduction Station, F.C. Bradford. And the Bradford pear took off. In the 1950s, it was planted on its own merit as a handsome, disease-resistant street tree to grace suburbs and cities. Impervious to pests and shrugging off most diseases, resplendently white-mantled in spring and ruby-leaved in fall, the Bradford pear seemed to shine.

But there were shadows lurking in the Bradford lineage. They took time to show: as late as 2005, one Callery cultivar was named the Urban Tree of the Year by the Society of Municipal Arborists.

And yet, the shadows were gathering. The Bradford was not as strong as it appeared. Time allowed us to see its weaknesses. It was a short-lived tree, rarely living over 30 years and prone to breakage in wind. Tight branch-to-trunk angles and congested branching practically invited the limbs to break apart under a snow-load. Replacing them became a regular chore for arborists around the country as they succumbed to injury and storms.

Worse, it was not as sterile as it seemed. While any one of the Callery subspecies was sterile when it was solitary, commercial plant breeders were coming out with all sorts of varieties of the

Callery pear, mainly as attempts to improve on the Bradford's poor branch structure and its unfortunate propensity to break. By the 1980s, a dozen were available for planting. Sterile by themselves, if the tree paired up with any other cultivar, it would set a viable crop of fruit that hungry birds gobbled down with glee.

Given that this is a very common pattern in fruit trees, with one set providing pollen and another producing the fruit, that really should have been a red flag. Spread by the digestive tracts of birds, these seeds were scattered throughout the Northwest and Midwest. But what rose from them wasn't the well-behaved and well-tended suburbanite Bradford pear. No. These were feral cross-bred children of all the dozen pear types, and they had reverted to their ancestry: arming themselves with thorns, growing in dense thickets, and quite happy to take a mile anywhere the native flora gave them even an inch. To make things worse, the mixed genes gave the trees blooms that smell like the fish that's left in the shop at the end of a long, hot day.

It's easy to wrinkle your nose at a Bradford pear. But don't blame the tree. It's what was done to it that has made it into the feral beast it is today. Our botanists brought it to America. Our horticulturists selected for a handful of traits, ignoring all others. Our contractors and homeowners planted it across the country, ignoring the warning signs. The Bradford pear is humanity's creation, a mixed-up wild child of a tree that belongs neither here nor there and will grow anywhere. They say you reap what you sow? We sowed the hybridized Bradford pear disaster. Now, we're reaping it.

And yet, for all its harms, there is something to be said for a Bradford blooming on a spring morning; the sunlight gleaming through white blossoms. There's a beauty in the gleam of red leaves on a crisp October afternoon.

Hubris. It's alluring. It lulls and charms us into thinking we're doing right. But like the Bradford, hubris has thorns. And it can tear your heart.

If You Love It: You can always admire the existing trees, but arborists, foresters and natural resources scientists strongly discourage planting Bradford pear trees, even if some nurseries continue to sell them. The trees have been banned in the state of Ohio and two American cities: Charlotte, North Carolina, and Pittsburg, Pennsylvania. Most states list it as an invasive species. Even if it charms you, please don't add to the problem.

If You Loathe It: I wouldn't blame you. Between thorns as long as your pinky, flowers that smell of stale fish, and a terrible tendency to muscle out every other plant, there are a lot of reasons to get rid of Bradford. They've earned the ignoble nickname 'tuna on a trunk' for their smell, and the thorns can be dangerous to animals and children. So what do you do?

For the Home: A young sapling is easily dug up from the roots. Dr. Rudy Pacumbaba, Extension Specialist at the Alabama Cooperative Extension System, suggests that a larger tree would have to be girdled, and at least an inch of bark removed right above the soil line. "Over about a year, the tree will die, but considering the rootstalk itself it is very resilient, you'll have to keep an eye on it," he states.

The mature trees can be removed by professional arborists and the stumps ground to prevent suckering. Currently there are a number of programs that will subsidize or cover the cost or removing the trees, and help with replanting something more suitable. These include the Clemson Bradford Pear Bounty in Clemson, South Carolina and the Callery Pear Buy-back in Missouri. To find out if you can get some help, reach out to your local agricultural extension; they'll be able to direct you.

For Larger Areas: Very young plants can be controlled by regular mowing. A Bush Hog or similar tool can take down adolescent stands of the trees. After that, regular passes with a riding lawn mower will keep them from getting very far.

Burdock, Lesser and Greater

Arctium minus, Arctium lappa

Common Names: lesser burdock, bardane, beggar's button, burdock, common burdock, small burdock, smaller burdock, wild burdock, wild rhubarb, personata, bardona, appa major, and clot-bur.

Family: Asteraceae
Genus: *Arctium*

Looks Like: a stoutly tap-rooted biennial that flourishes in warm seasons. Arrow-shaped leaves with wavy margins and hairy undersides grow up to a foot long, arranged close to the ground in the first year. A stalk 5 or 6 feet tall emerges in the second year, topped with purple flowers. Flowering is July-October. The fruiting bodies are covered in hooked barbs, forming the classic burrs we all know and dread.

Comes From: Eurasia. I know that's a wide area, but it's been found all over.

Likes To Live: Any sunny, open and reasonably watered area below 7,300 feet. Disturbed fields are a favorite, as are river and irrigation ditch banks.

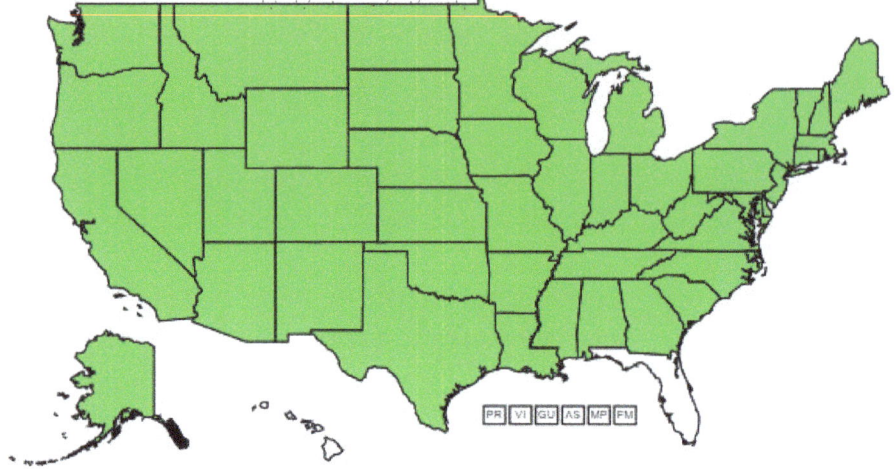

States Present: All 50.

States Impacted: Colorado, Wyoming, Kentucky, Pennsylvania, Delaware, Maryland.

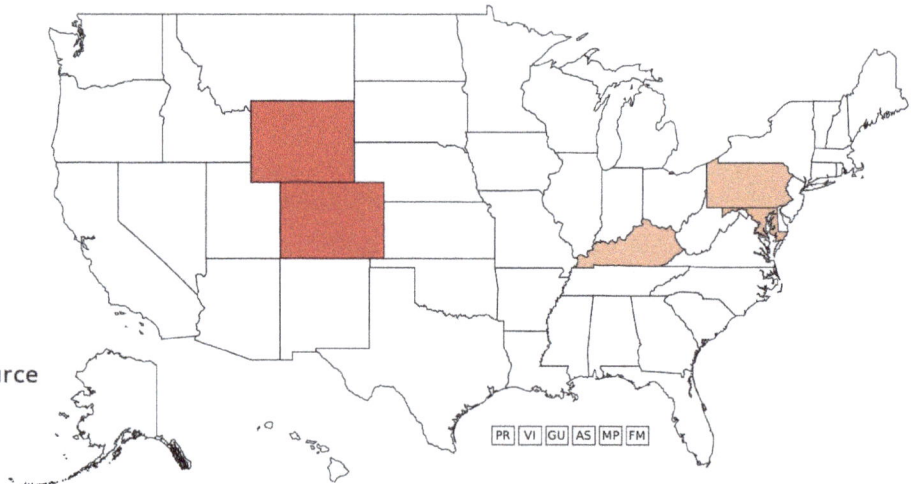

Legend
- List only
- Law only
- Both List and Law
- Included on other source
- Not Listed

26

Succeeds By: seeding prolifically. I mean, rapaciously. Rabbits have nothing on burdock. One plant typically produces 15,000 seeds in its flowering year.

Causes Trouble By: invading pastures, hay fields and open prairie ecosystems, where it becomes a pest. It acts as a secondary host for pathogens, such as powdery mildew and root rot, which affect crop plants. It reduces the value of sheep's wool when its seed heads get tangled up in their curls. It taints milk if the animals eat more than a bit of it, leaving the product bitter and unpalatable. The burrs often cause damage to the eyes and mouths of livestock.

Burdock's arrival in the United states was first recorded in 1672 by John Josselyn, a sharp-eyed English visitor. In a pamphlet called *New England's Rarities Discovered*, he listed, "such Plants as have sprung up since the English Planted and kept Cattle in New-England," including common burdock. It's a well-named plant: Artos or Arktos is the Greek word for bear, and lappa means 'to grab ahold' So the plant's Latin name could be translated as 'the one that grabs ahold like a bear' or 'the one that grabs on and bites like a bear', which I can certainly understand.

Burdock probably arrived in seed stock brought by the settlers, mixed with what they intended to plant. But there is some evidence that it may have been intentionally planted as a 'tonic herb', a plant whose bitter taste improved digestion and appetite. William Darlington put it best in his *Agricultural Botany: An Enumeration and Description of Useful Plants and Weeds* when he wrote:

> "Every body knows this coarse homely weed, wherever it has gained admittance,—but every body does not take care to keep it in due subjection. One of the earliest and surest evidences of slovenly negligence, about a farm-yard, is the prevalence of huge Burdocks. The plant is considerably bitter; and the leaves are a favorite external application in fevers, head-ache, &c."

It's definitely true that humans are one of the few creatures who will consume the plant voluntarily; burdock protects itself from predators with bitterly flavored leaves and with spiny hairs. However, there's always been a school of thought that something really unpleasant must be good for you, so burdock was often used to 'tone up' a patient. It was also believed that the plant was useful to break up kidney stones when drunk as a decoction, though this practice has died out in the modern era.

Burdock does have uses around the home and garden. First-year roots and second-year stems can be peeled and boiled for about 20 minutes, to be eaten as a wild spinach. Immature flower stalks can also be harvested in late spring; their taste is a bit like artichoke. In the compost heap (as long as you remove the burrs) it adds a lot of nutrients. In her wonderful resource *The Complete Medicinal Herbal,* Penelope Ody recommends a decoction of the root for external application to dry eczema, a tisane of the root used as an acne wash, and a half a cup of the tea made from the leaves before meals to ease digestion.

If You Love It: The plant has its medicinal uses, and to the right eye it has its charm. It's also a great winter feed for birds. The seeds should be stratified in order to germinate; they germinate well when sown directly in spring, after the danger of frost has passed. Plant the seeds 1/8 inch under the soil, and keep it evenly moist. Germination takes about 1 to 2 weeks. The young plants grow quickly, but it takes some time to establish a taproot worth harvesting. Just make sure you keep it from seeding, and don't let it wander into your neighbor's yard and misbehave.

If You Loathe It: I wouldn't blame you. The burrs are a disaster! To get the burrs out of hair (yours or otherwise), stop pulling. Instead, put on a handful of conditioner or vegetable oil, carefully working it into the hair. The slippery stuff will make it much easier to work the spines free. This also works for your pet's fur. To get burrs out of clothes, wash them in hot water and tumble them in the dryer. Lay the items out and run a fine-tooth comb over them. That generally does the trick. Never ever pull on the burrs; that will make them lock in.

Keeping burdock out of your garden is an awful lot easier than reclaiming land from it. To do this, keep your planting beds well mulched, pull any upstarts as soon as you spot them.

For The Home: If you're trying to reclaim a neglected garden, chop the plant to the ground with shears every time you see it to prevent them from seeding, and remove their old growths. You will need to do this for several years, until the seed reservoir in the soil is exhausted. The silver lining is that, as a biennial, burdock can be worn down fairly fast.

For Larger Areas: To reclaim a large space or maintain a right of way, mow or weed-wack any large areas of burdock several times in the season. Keep up very regular maintenance for at least three years.

The Chemical Option: Due to the size and nature of burdock, herbicides aren't a great idea. They grow too big too fast to make spraying efficient, and they tend to be too close to waterways for it to be safe. For this plant, a weedwacker is your best friend.

Canada Thistle

Cirsium arvense

Common Names: Canada thistle, Californian thistle, creeping thistle, corn thistle, field thistle, curse thistle, way-thistle, sting-needle, Hell's lettuce.

Family: Asteraceae
Genus: *Cirsium*

Looks Like: a 3 to 5 foot tall, lanky plant with oblong, spiny, bright green leaves that are slightly hairy on the undersurface. Canada thistle flowers form in small clusters of 1 to 5. They are about 1 cm in diameter, tubular, and vary from white to purple in color. Seeds are around a 16th of an inch long, with tufts of bristles that help them fly.

Comes From: the Eastern Mediterranean region. It was probably one of the first weeds that early settlers imported to North America. Oops…

Likes To Live: wherever it can, really. It'll colonize crop fields, pastures, rangeland, roadsides and abandoned places. Generally, it gets a start on disturbed ground like ditch banks, overgrazed pastures, tilled fields or abandoned lots.

States Present: It's easier to list where it isn't. The only states in the U.S. that are relatively free of Canada thistle are Texas, Oklahoma, Louisiana, Mississippi, Alabama, Georgia, South Carolina and Florida.

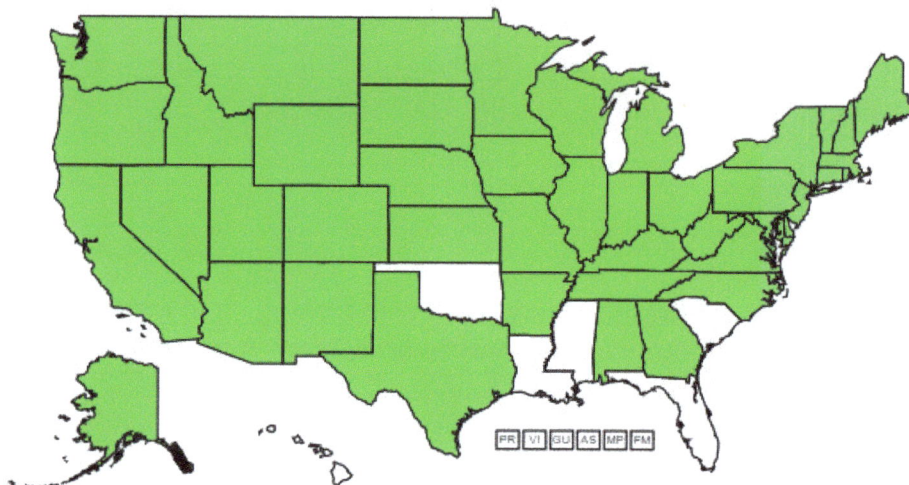

States Impacted: It's a particular problem in the upper Midwest, the Great Plains states, and the Pacific Northwest.

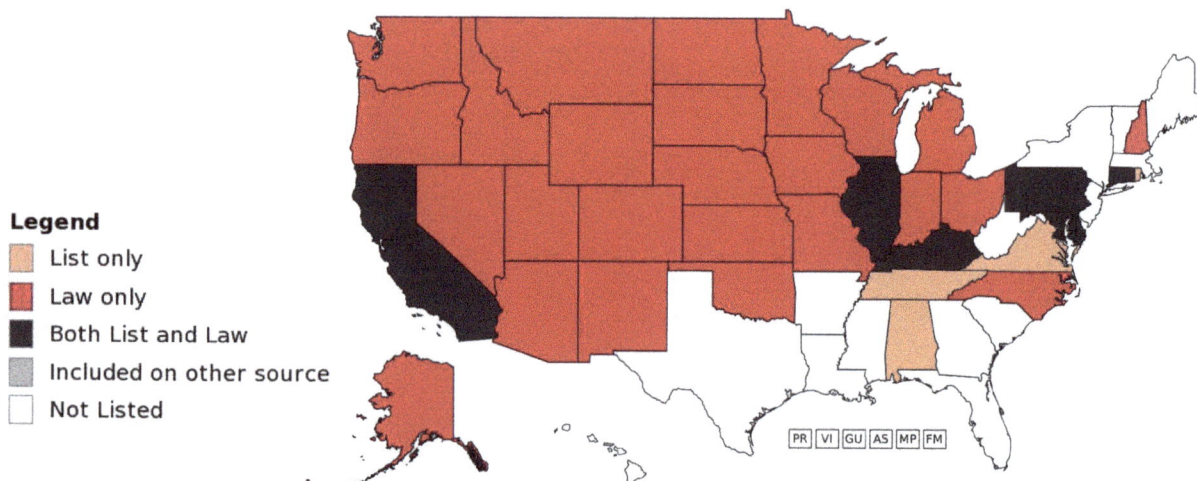

Legend
- List only
- Law only
- Both List and Law
- Included on other source
- Not Listed

Succeeds By: spreading like mad by seed, and growing incredibly fast underground as well. Each plant produces 1,500 seeds on average and may produce up to 5,300 seeds in a season! But the real key to Canada thistle's weediness is its root system. The roots of Canada thistle spread like wildfire, increasing the width of a thistle patch 6 to 10 feet in a season! As the root system spreads, it gives rise to new shoots. If it isn't controlled, a single Canada thistle plant will eventually become a patch containing thousands of stems.

Causes Trouble By: crowding out native species, reducing crop and forage yields. A density of 20 Canada thistle shoots per square meter causes estimated yield losses of 34% in barley, 26% in canola, 36% in winter wheat, and 48% in alfalfa seed. Canada thistle can form dense stands that shade out and displace native plants as well, changing the plant community and reducing biodiversity. It spreads rapidly and is very difficult to remove.

Canada thistle has a dubious honor in America: it is the first herbaceous plant to be branded an outlaw and an invader in the New World. "Be it enacted by the General Assembly of Iowa," read the words laid down in 1868, "that if any resident owner of any land in this state after having been notified in writing of the presence of Canada thistles on his or her premises, shall permit them or any part of the root to blossom or mature, he or she shall be liable to a fine of five dollars and cost of collection for each offense".

Given how Americans tend to react to laws that infringe on private property, you can imagine how this law went over. In fact, this frustrating plant has been the cause of arguments as prickly as its leaves. Take its most common names: Canada thistle, or Canadian thistle. The plant definitely isn't from Canada, so why the name?

Well, it's called Canada thistle because people like to fight. The plant was called "Canada" Thistle because early New England residents blamed its emergence on French traders from New France, their much despised neighbors. The dislike between the two colonies had started right from the beginning. About the same time John Smith and the Jamestown settlers were setting up camp in Virginia, France was building permanent settlements of their own. And the differences between the two colonies were stark. Unlike the English colonies, where self-rule had been instituted immediately, the people of New France had absolute allegiance to the laws of home. There were no elected assemblies. Decisions were made by local magistrates on behalf of the French king. Oh, and one other thing: they were strictly Catholic. Louis XIV was a devout man and a religious purist, tolerating no other faiths within the French Empire.

The differences between the neighbors could not have been more sharply drawn. The English colonists had local rule via assembly, informed and advised by the King, of course, but were expected to figure things out locally. They were, for their time period, supportive of secular approaches, and welcoming of all varieties of—I have to stress this—Christianity. By our modern sights, they weren't what we'd call a tolerant society, but for their time period and culture, the English colony was practically inclusive. So the relationship between these two approaches to life and rule were a bit… strained.

Quebec and Jamestown blamed one another for everything: plagues, sunken ships, even black magic. One of the few connections between the communities was the coureurs des bois: itinerant, unlicenced fur traders from New France. They were known as "wood-runners" to the English, and they traded up and down the land, often working with the peoples of the First Nations as well. Of course, that made the English suspicious. English settlers often accused the coureurs des bois of being what we would now call terrorists, accusing them of stealing children, livestock, and—yep!—sowing the seeds of unpleasant weeds in farm fields to bedevil English farmers. Hence, 'Canada' Thistle was blamed on the sneaky Canadian French by the English settlers. However, historians believe the plant appeared in the United States and Canada at approximately the same time. So how did it do that?

Honestly, both groups probably planted it in their fields, unwittingly. At the time, checks on seed were rudimentary at best. Farm seed shipments of the time arrived laden with exotic weed species; Canada thistle was definitely in there. Let that be a lesson to those who would blame their neighbors for their misfortune!

If You Love It: Canada thistle does have its charms, and it does feed birds and bees to some extent. But it's no friend to the native plant ecosystems, and I wouldn't recommend giving it any space anywhere you're caring for. Don't worry about taking food from the birds when you eradicate the plant either: those thistle seeds you buy for birds in the store are actually the seeds of the Golden Daisy Flower, *Guaroztica*. If you admire the form of thistles, why not plant or encourage native American species? The Xerces Society has some great thoughts on this:

> "Native thistles (genus *Cirsium*) fill important niches in our ecosystems. In great grasslands and prairies, alpine meadows, and silty Midwestern river bottoms, the seeds of our native thistles help sustain enormous flocks of songbirds such as goldfinches and indigo buntings. The nectar of these plants fills the stomachs of countless flower visitors, including the enormous black and gold bumble bee (*Bombus auricomus*), while the foliage of thistles feeds both people and rare butterflies alike. Edible thistle (*Cirsium edule*) is a staple food of the Salish people of the Pacific Northwest, while swamp thistle (*C. muticum*) is a caterpillar host plant for the swamp metalmark butterfly (*Calephelis muticum*), listed as endangered in Wisconsin."

If this interests you, I recommend reading more at the Xerces Society page, https://www.ecolandscaping.org/10/designing-ecological-landscapes/native-plants/bringing-back-native-thistles/

I'd love to see more people planting native thistles, which do have lovely form and make great dried flower arrangement pieces, as well as helping the ecology of your area. That said, even if you find Canada thistle's form charming, don't give this particular thistle an inch. It'll take the yard.

If You Loathe It: You definitely wouldn't be alone! I hate to tell you this, but you're in for a long hard row to hoe if you want to get the stuff out of your property.

For the Home: Shovel it. Dig as deep as you can, and get the taproot out. This isn't going to be fun, but it's really the best control method. Make sure you get the fan of running roots around it out too!

If the flowers are seeding, put paper bags over them, or put the entire plant in a garbage bag. You don't want to walk across your property spreading seed as you go.

Finally, sharp shoot it. There is one smart way to use systemic herbicide, preferably early in the season. Cut the tops off the plant, and then apply 2-4 D directly to the cut roots. You can use glyphosate if you must, but let's move away from that.

Now that you've taken action, continue to keep an eye out, and get after every spiky leaf that rears its head. You'll have to do this for 2 to 5 years, but eventually you'll wear the root system out.

For Larger Areas: For larger rangelands and large properties, really the best approach is going over the stands with a riding lawnmower twice a month until you stop seeing it. This will take several years, but it's the safest and most effective method. The only other method is to plow to the depth of the roots, but unless you immediately plant a good cover crop like alfalfa, vetch, or wildflowers, you're simply creating a nice prepared bed for all sorts of weeds. Remember, weeds are opportunists. You have to move fast to keep ahead of them!

The Chemical Option: Unless the thistle is extremely young, pesticides are worse than useless on Canada thistle: given the size and spread, you'd end up killing everything in the yard to get enough poison on the pesky thistle. Don't do that.

Couch Grass

Elymus repens

Common Names: Bermuda grass, devil's grass, knot-grass, quick grass, quack grass, quitch grass, scutch-grass, twitch, wheat grass. The name 'quack' is a variation of the German 'quecke', which means 'to live', referring to the persistent nature of this weed.

Family: Poaceae
Genus: *Agropyron*

Looks Like: a creeping, clump-forming perennial grass with straw-colored, sharp-tipped rhizomes and a pair of auricles (little structures like a bird's foot) that clasp the stem at the top of the sheath. It can form large mats over time. The leaves are blue-green and around 1-1/2 to 8 inches long (sometimes up to 12 inches) and 1/8 to 1/4 inch wide (sometimes up to 1/2 inch). The flowers are arranged in a long, slender, unbranched spike from 2 to 10 inches, resembling a slender head of wheat. It is in flower any time from spring to fall.

Comes From: Probably sub-Saharan Africa and/or islands in the western parts of the Indian Ocean, but nobody is completely sure.

Likes To Live: anywhere with a bit of water. It's happy with dappled shade or full sun, in compacted soil or in brackish conditions.

States Present: All of them!

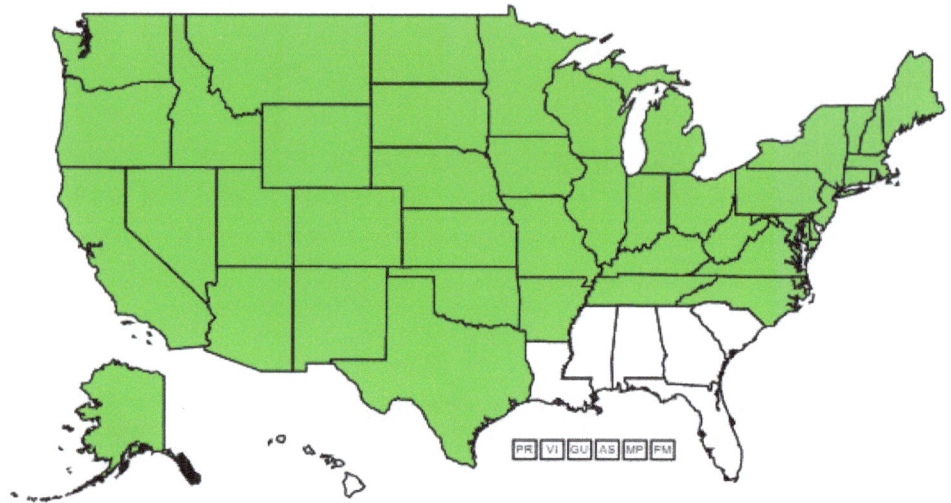

States Impacted: Listed as a noxious invasive in Alaska, Arizona, Hawaii, Iowa, Kansas, Oregon, Utah, Wyoming.

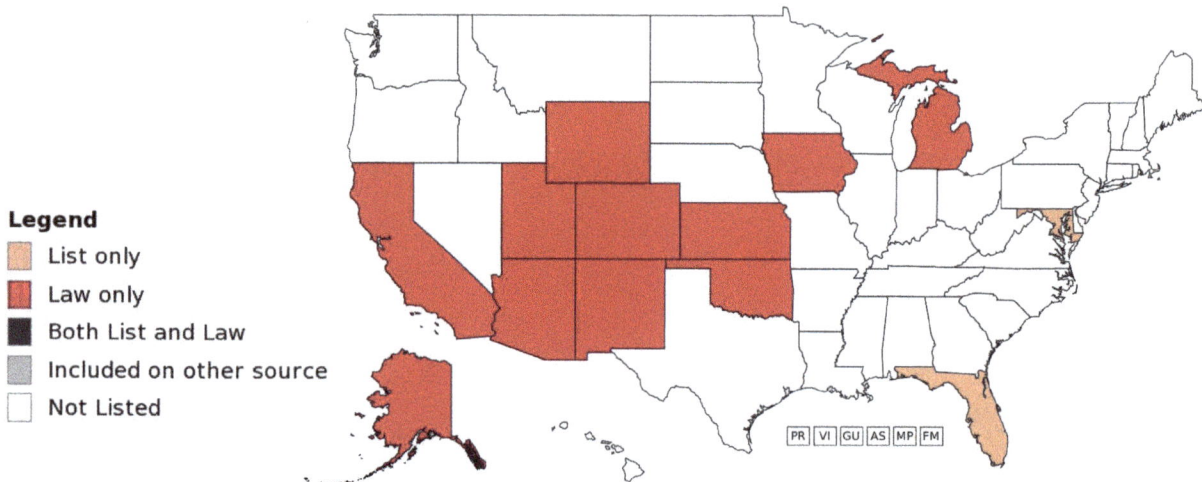

Legend
- List only
- Law only
- Both List and Law
- Included on other source
- Not Listed

Succeeds By: spreading like mad. Rhizomes of quackgrass can grow 6 to 8 feet deep in the ground, and these rhizomes push new plants to the surface when the mother plant is yanked out of the ground. Each rhizome has hairy little joints every few inches, and each joint can grow into a new plant. One plant can produce 300 hundred feet of rhizomes in a year! It also spreads by wind-blown seeds.

Causes Trouble By: outcompeting desired species. It can cause yield reductions (up to 85%) in crops like corn, and can also result in delayed corn maturity. It is an alternate food source for several insect and disease pests of grain, including the cereal leaf beetle (Oulema melanopa) and bromegrass mosaic virus. That makes it a bit more dangerous than your average weed to crop plants. In residential areas, it often outcompetes ornamental perennials and turf.

> "The ubiquitous Quack-Grass or Witch-Grass (with scores of other colloquial names) is usually known merely as a persistent and obnoxious weed, most difficult to eradicate and completely eating up the good of the land; but it was shown in the 18th century that it might be eaten if one cared to utilize it."

This charming little entry in *Edible Wild Plants of Eastern North America,* penned by the renowned botanist Merritt Fernald in 1943, tells us that the poor little couch-grass hasn't been much loved in its time in the United States.

But it was apparently valued in earlier days; in the little book from 1672 with the long-winded name *The American physitian, or, A treatise of the roots, plants, trees, shrubs, fruit, herbs, &c. growing in the English plantations in America : describing the place, time, names, kindes, temperature, vertues and uses of them, either for diet, physick, &c.,* a Mr. William Hughes tells us that it was introduced into New England by the colonists for 'forage and pottage'. That's not an easily accessible phrase to modern readers, but what was meant was: food for animals and seed crops that could be made into a form of porridge. The starch-rich roots and the oily seeds were parched together and ground into a fine meal called pinole, made into cakes and porridge. The grass would grow in areas that were heavily trampled by cattle, which made it ideal for the early settlers.

Well, for a while...

If You Love It: Couch grass does have a lot of value as a foraged food. The roots are mildly sweet and very fibrous. The long rhizomes can be cleaned, dried and ground into flour, or brewed as a coffee substitute. Raw rhizomes can be chewed as a quick garden snack. And it's good for you too: studies have shown that the seeds contains 12% oil, mucilage (10%); saponins; sugar alcohols (mannitol, inositol, 2% to 3%); essential oil with polyacetylenes or carvone (0.01% to 0.05%); small amounts of vanilloside (vanillin monoglucoside), vanillin, and phenolcarboxylic acids; silicic acid; and silicates. In the past, it was used for digestive and urinary disorders, and as a spring tonic. Given that it contains phenolcarboxylic acids, that's probably accurate. Just be careful that the place you get it isn't sprayed with herbicides!

If You Loathe It: You wouldn't be the only one. All sorts of things have been tried to kill couch grass, but in the end you really only have four viable possibilities:

Manage It: Couch grass really only has the advantage when the grass around it isn't doing well. So, attend to your grass. Get it a little more nitrogen—about 0.25 to 0.5 pounds nitrogen per 1,000 square feet every two weeks during the growing season—and a little more water, and the cultivated lawn will choke the quackgrass.

Handle It: If you haven't gotten the results you want by helping the lawn, the next step is getting your hands dirty. Be careful; if you simply cut into the plant, you'll be making more couch grass and more trouble for yourself. Instead, try this:
- Pour boiling water on the plant in appropriate areas
- Burn the top of the plant with a fire wand
- Cover what's left in mulch to keep it from getting vital sunlight
- If you do want to try digging it up, loosen clumps with a pitchfork and throw them in the trash
- Over large areas, tilling is your best option. Repeated shallow tilling exposes large numbers of rhizomes and forces the ones that remain buried to use up their food reserves.

Spray It: If you're at your wits' end, glyphosate can kill it. But then, glyphosate kills everything, doesn't it? If you choose this route, you will need to apply the chemical twice. Make the second application 14 days after the first. Seven days after the second application you can rough up the area and sow your new grass seed.

Accept It: After all, couch grass has many worthwhile qualities. The roots form a dense mat, holding soil in place, and the plant can weather both salinity and compacted soil. And it's very good at staying green. On an abused, compacted hillside, maybe let it grow? Kept mowed, it almost blends in with other grasses… almost.

Crabgrass

Digitaria sanguinalis

Common Names: Crabgrass, finger-grass, rabbit crabgrass, crab-finger grass, purple grass, po, red fingers, manna grits, fonio, crop grass, kasha.

Family: Poaceae
Genus: *Digitaria*

Looks Like: a coarse, flat-leafed summer annual grass, with blades measuring up to 6 inches long and up to 1/2 inch wide. Young leaves unfurl from tight spirals. The mature leaves have a prominent mid-rib and hair on both sides. There is often a deep red or purple streak along the midribs, which earned it the Latin appellation it bears. 759, Lorenz Heister named the genus *digitus*—Latin for finger—referring to the plant's finger-like inflorescence. The name *sanguinalis* refers to the plant's blood-red or purple streak. The growth form is prostrate, and the plant often forms a circular mat that can grow to 15 to 36 inches in diameter when it's not mowed. Flowering takes place June through October. Flowers cluster along 3 to 7 slender, fingerlike branches that are located toward the end of the flowering stem, yellow-green in color.

Comes From: Europe and Asia.

Likes To Live: anywhere the ground has been disturbed. Weedy meadows, edges of degraded wetlands, areas along roads and railroads, lawns and gardens, vacant lots, fields, grassy paths, and sundry waste areas. It has a preference for full or partial sun and heavy clay-loam soil.

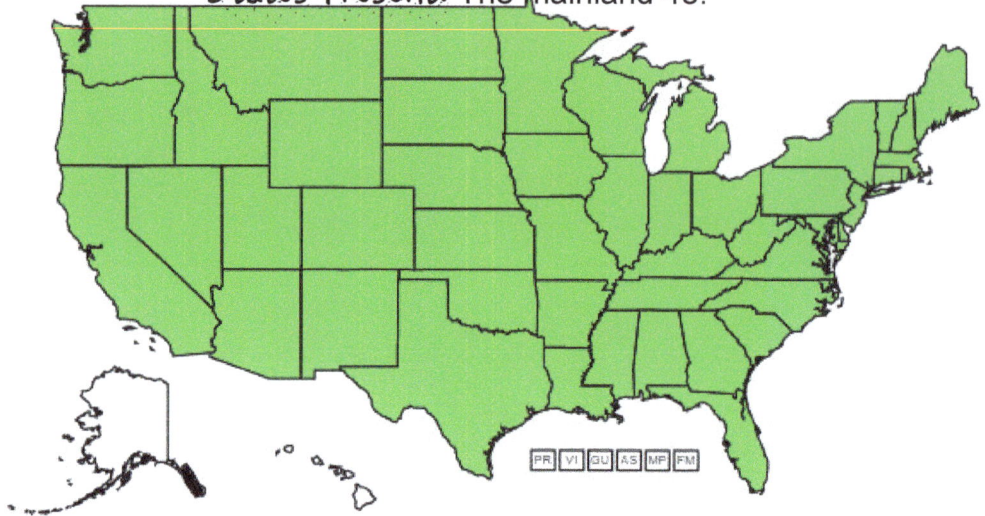

States Present: The mainland 48.

States Impacted: The states of the northeast and the south, when it invades crop fields. It is:
- One of three most serious weeds in peanuts
- A principle weed of sugarcane, cotton, and sorghum
- The most prevalent weed in spring alfalfa seedlings
- A major weed of turfgrass, particularly in the Southern U.S.
- A major turfgrass weed throughout the U.S.

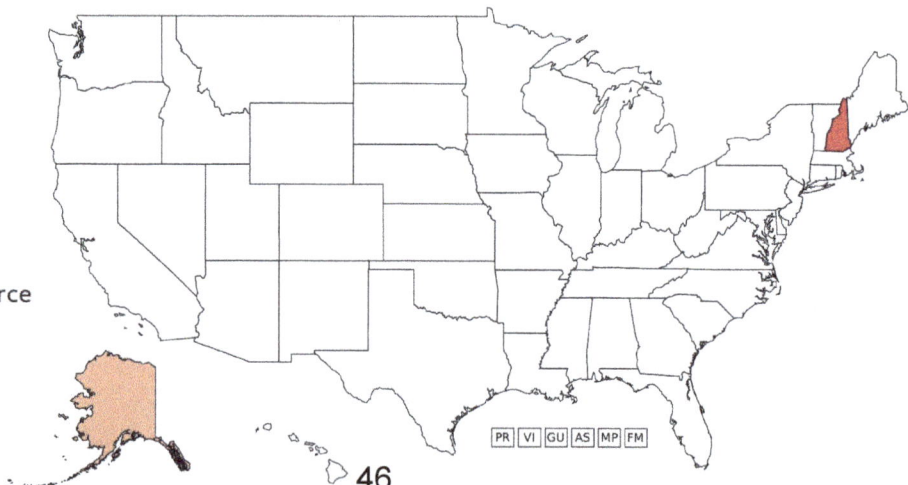

Legend
- List only
- Law only
- Both List and Law
- Included on other source
- Not Listed

46

Succeeds By: growing fast, being tough, and spreading via both rhizomes and copious seed production. A single plant is capable of producing 150,000 seeds a season. The seeds survive ingestion with no trouble, and can spread in the manure of animals. Only cold really slows it down; it prefers to germinate near the surface, when the soil temperature is warming up. When the soil temperature at the surface reaches 55°F for four or five consecutive days, crabgrass begins to germinate. And that's when the trouble starts.

Causes Trouble By: acting like a bully on the playground, muscling any plant that is weak out of the way and taking up the space. This makes it an impactful plant in both turf areas and in food crops. In several crops, it lowers quality and yield. Its fast growth in hay meadows contributes to stand-thinning from spring to late summer.

Crabgrass also has a slower drying rate than most hay species, which causes rotting and mold development after baling. It can outcompete peanuts and alfalfa, and act as a nutrient thief among taller species such as sugarcane and cotton. Crabgrass is especially competitive for nitrogen and potassium, and can deplete the soils. It may also act as an alternative host to crop pests and diseases, including rice stripe virus and rice black-streaked dwarf virus, Pangola stunt virus of sugarcane, sugarcane mosaic virus, maize streak virus and sorghum plantbug.

It may be lowly in these modern days, but crabgrass has a distinguished history. Crabgrasses were among the first cultivated grains, dated back to 2700 BCE in the archeological sites of China. Crabgrass has been grown for food in India and Africa, where it is known as *fonio*, and it was a staple food in several ancient communities of Switzerland. In the well-preserved Stone Age pile communities of the Three Lakes Region, crabgrass seeds were found in storage pits.

But why crabgrass? Lots of reasons! Crabgrass is in the same family as millet, and it's actually a great source of nutrition. Not only is it nutritious, but it's one of the world's fastest growing cereals, producing edible seeds in six to eight weeks. How were they eaten? Lots of ways. Crabgrass seed can be used as a flour, couscous or as a grain, eaten as a porridge or fermented for use in beer making. And—always a favorite—it can be made into meat and dairy by pasturing animals on the nutritious grass.

Even its name tells the story: if you trace it back, it was originally 'crop-grass', not 'crab-grass'. Sources as late as 1815 use this older and more complimentary name; it looks like the moniker was mispronounced into its current form here in the USA. Resilient, energy-dense crabgrass makes for some very happy cattle, sheep, and goats when it's used as a forage crop. The seeds are suitable for fattening hogs. It grows well in dry areas with poor soils, and flourishes with a little water. And it couldn't care less about being trampled. It's the perfect forage crop, especially in the Southern United States.

In fact, that's what brought this tenacious little beastie to the New World. According to the Weed Science Society of America,

> "Immigrants from Poland and central Europe in the late 19th century and early 20th century brought crop and garden seeds, including crabgrass, with them. They held the crop in high esteem, calling it "manna grits." A form of millet native to central Europe, crabgrass thrived in many soil types, grew and matured rapidly, and produced abundant yields, making it a reliable crop. The United States Patent Office introduced large crabgrass into this country as a forage crop in 1849, a time when livestock numbers were increasing rapidly and good forage was scarce. The Department of Agriculture, founded in 1862, later assumed plant introductions. Decades later when southern ranchers sought a good forage grass for the expanding cattle business, they found large crabgrass perfectly suited because of its affinity for hot, dry conditions and its nutritious qualities as pasture and hay. Immigrant farmers eventually abandoned crabgrass in favor of wheat (*Triticum aestivum* L.) and corn (*Zea mays* L.) which were more profitable and easier to grow."

And that, my friends, is when the trouble began. Abandoned as a crop, crabgrass was left to its own devices. Escaping to any disturbed ground it could find, crabgrass colonized just as voraciously as the people who had brought it. The same qualities that had made it such a staunch crop now rendered it a most pernicious weed. In 1914 Ada George wrote, "the seeds of this grass must be very long-lived, for, though it is never sown, let the ground be cultivated, and as a general thing Crab-grass will be there."

So the next time you cuss at crabgrass, don't glare at the plant. If you're of European descent, give that dirty look to your great-great-way-back grandpappy.

If You Love It: then you're in a wise and discerning minority. If you can plant it far enough away from other economically valuable crops, crabgrass makes excellent forage. The concentration of crude protein in crabgrass is sufficient to meet the requirements of dairy heifers, young goats, does, bucks, mature ewes, and finishing and replacement lambs.

If You Loathe It: then you're also in good company. *D. sanguinalis* is classified as the 11th worst weed in the world. It's the second most frequently reported weed in a 1968 survey of farmers. Suffice it to say, a lot of people want to control crabgrass. To do it, here's some tips.

For the Home

- **Keep it down:** Mowing in household settings will keep the plant under control and prevent seeding. Beating the grass to the seeding stage is key.
- **Keep after it:** The key of crabgrass control is keeping it from gaining a foothold. Watch out as soon as the temperature goes above 55 degrees F, and yank every crabgrass sprout you find. You won't beat crabgrass in a season, not with the seeds it produces. But several years of concerted effort will get it under control and cut down on the number of seeds waiting to sprout.
- **Keep your garden mulched:** Since crabgrass needs to be near the surface of the soil to germinate, a layer of good mulch at least four inches thick will really help.
- **Keep your lawn healthy:** If you're not ready yet to take your grass lawn out and plant a mulched garden (which is what I recommend), the best protection against crabgrass is keeping the lawn thick and lush. That means watering deeply and infrequently. Don't sprinkle it often; this produces shallow roots vulnerable to crabgrass infestation. Also, a lawn should be mowed no shorter than 2 or 3 inches. Mowing shorter than that weakens the turf and gives opportunistic weeds like crabgrass an opening.

49

For Larger Areas: Browsing in crop land will keep the plant under control and prevent seeding.

The Chemical Option: Keep the pre-emergent handy. While most post-emergent herbicides are pointless on crabgrass, and glyphosate is worse than useless—it's just going to kill everything in the area of spray—there are some effective pre-emergents in the case of very bad infestations.

Creeping Bellflower

Campanula rapunculoides

Common Names: Canadian harebell, rampion bellflower, creeping bellflower, rapion bellflower, rover bellflower, creeping bluebell, European bellflower, Korean bellflower, garden bluebell, June bell, rapunculus, rampion, old man's whiskers, Capitol Hill weed, Cap Hill weed.

Family: Campanulaceae
Genus: *Campanula*

Looks Like: a perennial with upright spikes, one to three feet high on a mostly smooth stem, usually not branched. Leaves of the upper stem are lance-shaped and alternate, with irregular teeth. The basal leaves are wider and have a rounded, heart-shaped base. There may be some sparse hair on the leaf, especially on the underside. The largest leaves can be 5 inches long and 2 inches wide. The flower is an upright spike. Flowers appear along one side, violet-blue and 5-parted, with a bell-shaped corolla that is slightly nodding. They are an inch long, with pointed tips that flare outward as the flower opens.

Blooming season is from July to September. The mature flower produces a 3-celled seed capsule. The capsule has several shiny, light brown ovoid seeds that have small wings or ridges on the side. The seeds are dispersed by the wind as the stem is shaken. It spreads by underground rhizomes and produces deep, taproot-shaped tubers. Both are white and fleshy. Because any piece of the roots can sprout into a new plant, it is extremely hard to eradicate.

Comes From: the Caucasus and Siberian regions of Europe.

Likes To Live: on roadsides, old gardens, railway lines, and any disturbed areas. It prefers some shade in the heat of summer, but it doesn't need much water.

41 out of 50: everywhere north of Texas is stuck with it!

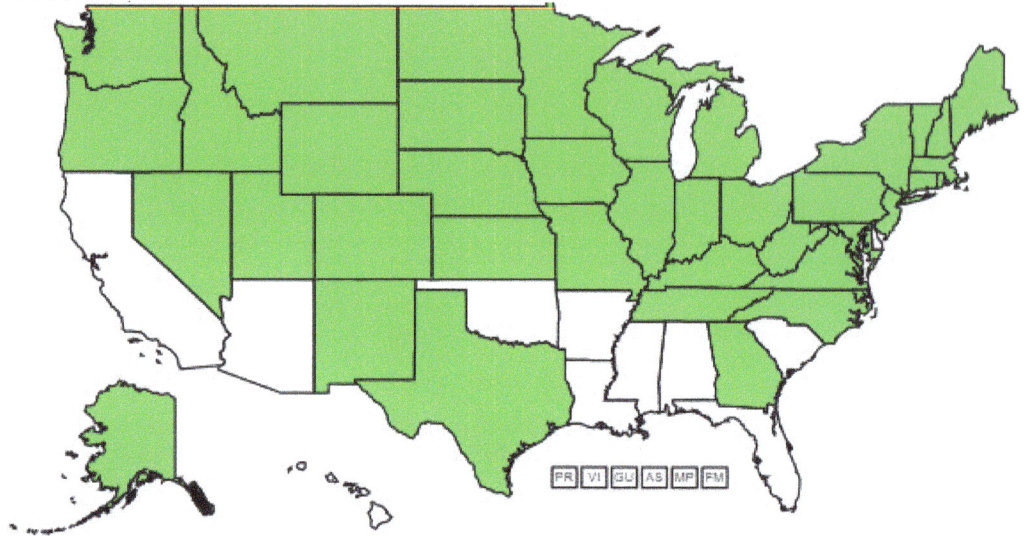

States Impacted: None directly, but this pretty stranger spells real trouble ecologically. Being extremely tough and creating monoculture stands, it takes over and depletes habitats of plant diversity.

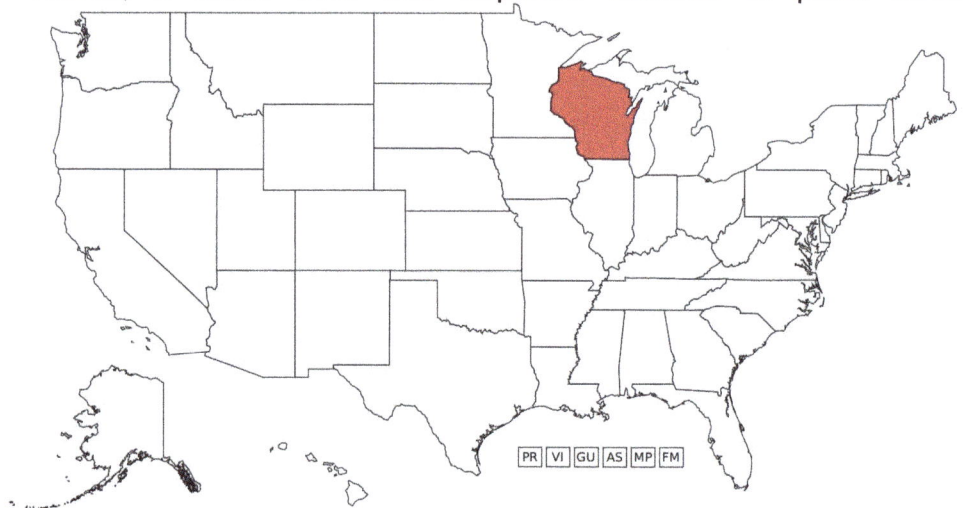

Legend
- List only
- Law only
- Both List and Law
- Included on other source
- Not Listed

54

Succeeds By: toughness, spreading rhizomes, and prolific seeds. It is estimated that a healthy plant can produce 15,000 seeds annually. The plant's creeping roots can travel under fences, lawns, and concrete, making it very difficult to control. Oh, and it's chemical-resistant to boot.

Causes Trouble By: being prolific and aggressive, outcompeting native species through high seed production, hardiness, and endless numbers of rhizomes. It's sneaky too: not only do its rhizomes slip into every crevice under the soil, it masquerades as several native Campanula species, such as Harebell. Some native violets have leaves which are so similar that people leave the invader alone until it's well into flowering, thinking it's one of these garden friends rather than a foe.

"But it's so pretty!" You hear it again and again when you point out creeping bellflower and explain that it needs to be eradicated. "But I like it!"

With its gently nodding mauve blooms and soft leaves reminiscent of an English woodland, creeping bellflower has gotten a lot of unsuspecting humans to like it… at first. It's charming, until it's been in the ground a year or two. Then it gets up to its usual tricks: spreading everywhere, choking everything else in the garden out, and generally being an insidious little beast. When you watch every other flower in your garden succumb under a blanket of purple blooms that become progressively more ratty as September draws in, those pretty blooms lose a lot of their charm. But there's always some naive human who says "oh, how pretty… can I have some to plant?"

And this is exactly how the lovely little troublemaker was given its foothold in the United States. Believe it or not, creeping bellflower was actually sold in the nursery trade right up into the 1930s. In my Capitol Hill neighborhood of Denver, its nickname is the Cap Hill Weed (many jokes about it being Capitol Hill's *other* weed, this being a permissive state where marijuana is legal.) Local lore says it was introduced in the 1920s to grace the grand old mansions that once stood in their majesty along these streets, and it's been with us ever since. It wanders from garden to garden in the older parts of town, sneaking under trees and into small areas, worming into bare spots in the grass and gaining a foothold most anywhere. Some people have simply accepted it as a garden plant like their grandmothers did, but most of us fight it tooth and nail because we'd like to have a garden with more than one plant in it!

Folklore winds around the plant like its thready white rhizomes. It is a classic part of the diet in Germanic and Caucasus countries, considered a favorite spring and summer green, and a quality root vegetable. This explains the second half of the plant's Latin name, *rapunculoides* meaning "like a little turnip".

In fact, rampion was so highly prized that a man in a legend once stole the plant from a witch's magic garden to help his wife in childbirth. The witch demanded the couple's baby in recompense, and imprisoned the girl in a tall tower. She named the girl Rapunzel, after the plant she was forfeited in trade for. Isn't that a surprise!

Interestingly enough, genetic and sociological studies show that the countries gifting us with the story of Rapunzel may have also given us the dubious gift of the eponymous flower as well. A paper on the genetics of campanula species published in *Molecular Phylogenetics and Evolution* makes this statement:

> "Biogeographic and dating analyses suggest that Western Asia and the Eastern Mediterranean region (Anatolia, Balkans) played a major role as centers of migration and diversification within the *Campanula* alliance. This is probably explained by the intense orogenic activity that took place in this region."

In more accessible terms, that means that the plant probably comes from the Balkan region. Comparing the date of arrival in the United States and the genetic divergence in the strains, we can trace the plant back to the port towns of the East Coast, probably in the 1840s. If we look at history, we see that it fits with a significant event: the first major wave of Czech immigration to the United States was going on in 1848, when the Czech 'Forty-Eighters', as they were nicknamed, fled to the United States, escaping the political persecution by the Austrian Habsburgs. As creeping bellflower was loved well enough to be on stamps in their native country, and an inoffensive flower and pot-herb, it must have seemed like a harmless thing to do... until it had been in the gardens a few years.

As the Czechs became backbones of the steelworking and coal mining community, creeping bellflower became a fixture in the gardens of the East Coast and Midwest. Without diseases or predators from home to hold it back, creeping bellflower crept out of gardens, spreading with the European immigrants and making its own way too. Soon, it had planted itself throughout North America. In 1870, it was recorded in the Canadian book *North American Wild Flowers*, where it's written up in these terms:

> "Although, in colour and shape of the blossom, the Canadian flower resembles the British one, it is more robust in its growth, less fragile—the flower stems being stouter, and the foot-stalk or pedicel stiffer and less pendulous, and yet sufficiently graceful."

Sufficiently graceful in appearance, yes... but given how fast it spread in only twenty years in order to be written up in Canada in 1870, I think its behavior lacks some elegance.

If You Love It: I suppose we've all had at least a moment of looking at these bell-like blossoms and thinking 'oh, aren't they nice', so I'm sure no-one will fault you (much) if you admit to a sneaking soft spot for the little woodland look-alike. It does have a number of uses in the home, both in the living room and the kitchen. If you're already saddled with a stand of the plant, it makes a decent cut flower.

And the good news is, you can eat it! And you should; as much as you can dig up. Like a lot of other greens you might forage, creeping bellflower is better cooked than it is raw. The leaves, even at their youngest, are a bit chewy; think celery, but less flavorful. You could probably hide a few leaves in a salad, but you're better off cooking them like spinach. On the other hand, the rhizomes and tubers are both nutty and wonderful raw, and the flowers are sweet; the perfect thing in a salad. All parts of the plant are rich in vitamin C. Oh, and if you have a handful of the rhizomes, try them for staking twine. They work beautifully.

All that said, nobody should be intentionally planting creeping bellflower. Shockingly enough, I am still able to find the seeds for sale on the internet. But please, I beg you: do not plant it. We have enough problems with it as it is. Treat creeping bellflower like a very mischievous pet: keep it tightly disciplined, preferably in a pot it can't get out of. Don't let it make a mess everywhere; cut the flower stems off before it seeds, every season. And never let it out of your sight! If you like the way it looks but want a less destructive plant, please choose *Campanula americanus,* or one of the native harebells. Your ecology will thank you!

If You Loathe It: Then you're in for some hard work.

For the Home: Mechanical removal seems to be the only feasible option for creeping bellflower removal. Grab your tools. Dig at least 6" deep to locate and pull all (and I do mean all!) the rhizomes and perennial roots. If you manage the amazing feat of getting all the rhizomes and tubers, stands can be eliminated. But this can be incredibly difficult, especially in heavy or hard-packed soils.

A pitchfork can be a lot of help for loosening the soil around the plant to make getting it out easier. Just be careful to break as few of the roots as possible; what stays in the ground will be a new plant soon. If you're only removing shoots—as you probably will be in an established garden you don't want to till up completely—it'll take a couple years of concentrated effort before you exhaust the nutrients stored in the roots. Keeping after it is the key.

Another Option: if you're lucky enough to be dealing with an area that doesn't already have plants you value in it, a spring burn works too! Spring burns can kill germinating seedlings and can suppress above-ground growth of established plants, depending on fire intensity. For small areas, a hand-held firewand can clean things up beautifully. Open burns are possible, but it's best to check with fire regulations in your area and plan ahead carefully. Be smart about this: make sure you're not near anything flammable, control the flame, and never do this on a windy day.

With safety in mind, I'll add that fire may benefit other well-adapted species (like prairie grasses), which can help them compete with bellflower. A handheld propane torch is my personal favorite tool for dealing with bellflower in the tricky spots such as flagstones, brick walkways and paving. Just give those leaves a quick burn, and they're controlled for another week. It won't be a fast chore, but the reward for having a diverse garden is worth the battle with the bell!

The Chemical Option: Herbicides are pretty much ineffective. Creeping bellflower is tolerant to many herbicides, including glyphosate. A paper from HortTechnology produced in 2018 states,

"Glyphosate and dicamba are often reported as options for chemical control of creeping bellflower, but repeated applications may be necessary for effective control. Estimated LD90 (the point at which 90% of the weed dies) of dicamba was outside the range of recommended rates for perennial weed control. Thus, field use rates of dicamba may not provide good control of creeping bellflower. Repeated applications may, however, improve creeping bellflower control by dicamba. In our experiments, glyphosate LD50 was more than three times greater in Expt. 1 than Expt. 2, but overall creeping bellflower would have been controlled (≥90% control) by glyphosate at field use rates. Repeated applications of glyphosate may be needed to control creeping bellflower.

Dandelion

Taraxacum officinale

Common Names: Pigsnout, fly-away, blowball, cankerwort, beggar's clock, shepherd's clock, clockflower, Irish daisy, piss-the-bed, pissinlit, priest's crown, puffball, yellow gowan, witch's gowan, dandy-lion, lion's tooth, bitterwort, monk's head, tarashaquq, tell-time, poor-man's-clock, wild endive.

Family: Asteraceae
Genus: *Taraxacum*

Looks Like: a perky little perennial growing from a thick, unbranching tap root. The deeply toothed, deep green leaves emerge from the crown of the plant in what is called 'basal growth', between 2 and 15 inches long and 1 to 4 inches wide. They're oblong or lanceolate in shape, with the bases gradually narrowing to the petiole. The bright yellow flowers grow on hollow stalks that may hug the ground or reach as high as six inches.

There can be 1 flower, or as many as 10, between 1 and 2 inches in diameter. These are light-sensitive and will close with the diminishing of light, at sunset or on grey days. The flower head is actually a tightly packed mass of many tiny florets, typical of the Asteraceae family. Each petal represents a single flower. The plants have a non-toxic milky latex that oozes out if leaves or stems are broken.

At maturity, the flowerhead dries into a white globe of loosely-attached seeds equipped with white parachutes, just waiting for the breeze. The seeds—called cypselae—range in color from olive-green or olive-brown to grayish, about 2–3 millimeters long with slender beaks. The fruits have 4 to 12 ribs. The silky pappi, which form the parachutes, are white and around 6 millimeters wide. Books will sometimes tell you that bloom time is spring, sometimes summer: in my experience, it blooms whenever it's feeling fairly healthy, from soil thaw until hard frost in the fall.

Comes From: Eurasia. Fossil seeds of Taraxacum have been recorded from the Pliocene of Southern Russia; this may be its origin point.

Likes To Live: anywhere it can get sun and a spot of water. You'll see it growing in cracks in the sidewalk, seemingly sprouting from the concrete itself. Give dandelion half an inch, and it'll take the yard!

States Present: Every single one. Our Canadian neighbors are having trouble with it too.

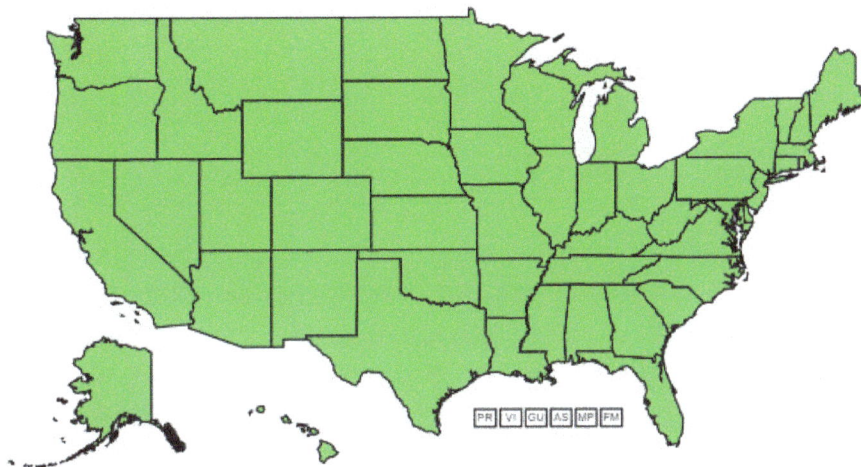

States Impacted: Pretty much all of them. But most have given up listing it.

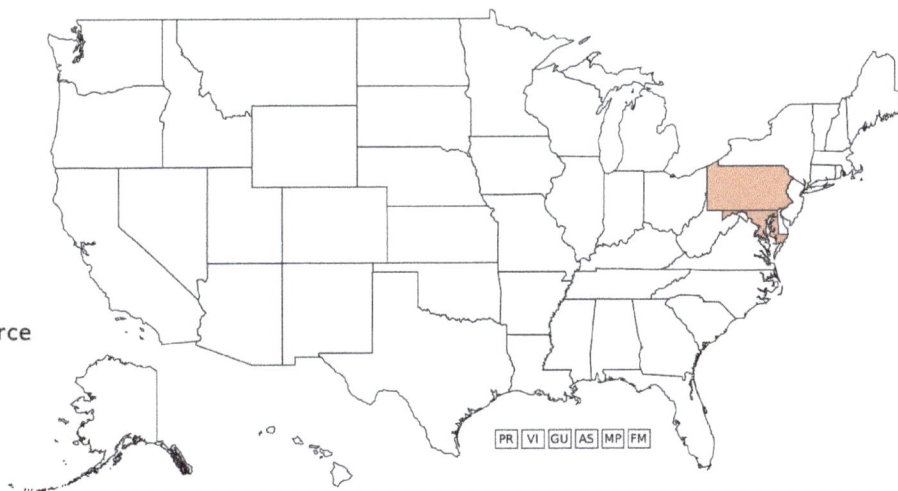

Legend
- List only
- Law only
- Both List and Law
- Included on other source
- Not Listed

Succeeds By: SEEDS! Seeds, everywhere, all the time, floating on the wind. There are usually 150 seeds per flower, and up to 10 flowers per plant. The plant usually blooms ten times in a season. How many is that? It's 15,000 seeds. Per plant. Per growing season. Dandelion florets also possess both male (pollen-producing) and female (seed-producing) parts. The fruits are mostly produced by asexual reproduction, without fertilization. This means that a lone dandelion is perfectly capable of producing viable offspring all on its own. When you know all that, the number of dandelions you see around the place makes a lot more sense.

The seeds can be dispersed across continents, because their pappi float them along in updrafts. And once the seeds are in the ground, they're no slouches, either. When a dandelion seed germinates, it produces a rosette of tiny leaves and concentrates most of its energies on growing a deep tap root. By the time you see a dandelion plant, it's already deeply rooted and well-established. The dandelion taproot can dig itself up to fifteen feet into the soil, and can regenerate itself from a half-inch piece of root when the aerial parts are removed. No wonder it's such a challenge!

Causes Trouble By: acting as a vector for disease, effectively competing for nutrients with desirable plants, and generally degrading a site's aesthetic appeal. It can be damaging to high-grade sports turf. That's not such a big deal in the scheme of things; mainly it bothers folks who could use a reminder to let nature be a little more wild.

But it's another story when dandelion gets into crop fields. It's becoming a problem in annual cereal and oilseed crops in western Canada, and was ranked the 6th most important weed occurring in corn, soybean and winter wheat fields in southwestern Ontario by Frick and Thomas in 1992. It was the 6th most abundant weed species in reduced and no-till fields, and the 10th most abundant species in conventionally-tilled fields. In the USA, corn yields were drastically impacted, especially in fields where there isn't regular crop rotation. It can compete seriously with seed-grown crops, especially beets, oilseed rape, lettuce, soyabeans and other beans.

It's rarely a problem in pastures and forage, as animals will happily eat it. It is a problem in hay, though; it can cause slower drying and moulding in the hay crop because of its high water content, and it can cause non-productive spots in the pasture where its leaves are shading out the ground.

More insidiously, dandelion provides an alternative host to several important viruses. These include tomato ringspot virus, spread by the American dagger nematode (*Xiphinema americanum*) the tomato-spotted wilt virus and the cucumber mosaic virus. Together, these infect many agricultural, ornamental and greenhouse crops. It may also act as an alternative host for boll weevils, cabbage looper, yellow-striped armyworm, green peach aphids, and larvae of the apple moth (*Lacanobia subjuncta*). There's some proof that it hosts bacterial streak and bulb rot, which infects onions.

When you say 'weed', it's the dandelion that springs to mind. It's the quintessential weed: tenacious, cheeky and indomitable, getting everywhere and into everything. It'll drive you mad some days, and on other days the shining yellow blooms will raise your spirits and remind you that you're living in a beautiful world. It's impossible not to smile when a child presents you with a bouquet of dandelions; being young, they've yet to learn which flowers are 'weeds'. All they see is the shining yellow bloom smiling up at them. And our children are more right than you'd suspect.

Dandelion is a plant with a very long history in humanity's gardens and medicine cabinets. The earliest western record of the plant was a Latin translation of the Arabic texts in 1170, by Gerard of Cremona. Its Latin name, *Taraxacum*, is a Latinized version of the Turkish name for the plant: tarashaquq, 'bitter pot-herb'.

It's recorded in pharmacological texts by the scientist Al-Razi, circa 900, and later by scientist/philosopher Ibn Sīnā (known in the West as Avicenna). Two centuries before that, Chinese medical texts were recommen ding Po Gong Ying for treating poor lactation, as a diuretic and as a liver stimulant. It also could be named from the Greek word taraxos, "disorder" combined with akos, "remedy." Theophatrus (371–287 BCE), the Greek philosopher and pupil of Aristotle, wrote of the dandelion in his *Enquiry into Plants*.

Folklore tells us that the Roman Army planted dandelions in every country they invaded, to feed their troops. Before any of them, the ancient Egyptians recorded their uses for the little plant in the *Ebers Papyrus*, circa 1550 BCE. That gives you a sense of just how long this plant has been aiding humanity. And our ancestors had many reasons to value the sprightly little perennial. It was often described as a 'tonic', a 'blood-cleanser', and a 'leavener'. And no wonder; it turns out that dandelion is a plant stuffed full of vitamins that humans suffer quite a lot without. Ancient physicians didn't know much about nutrition and vitamin deficiencies, but they knew for a fact that dandelions helped with a host of problems: kidney, stomach and liver disorders, skin irritations, heartburn, gallbladder problems, diabetes, arthritis, anemia, constipation, 'melancholy', hangovers, fevers, scurvy, and even warts! The detailed medical study "The Physiological Effects of Dandelion (Taraxacum Officinale) in Type 2 Diabetes" also encourages the use of dandelions in the diet, especially for diabetic conditions:

> "Dandelion has been considered a key anti-diabetic plant because of its anti-hyperglycemic, anti-oxidative, and anti-inflammatory properties. This is due to the various bioactive components present in dandelion, including polyphenolics, sesquiterpenes, triterpenes, and phytosterols. The most important and comprehensively studied bioactive components of dandelion are chlorogenic acid (CGA), chicory acid (CRA), taraxasterol (TS), and sesquiterpene lactones (SEL). These components possess great potential as anti-diabetic pharmaceuticals and nutraceuticals for regulating diabetes. They also have potential for use in anti-diabetic functional food."

Ounce for ounce, dandelion is incredibly nutritious as a food. The leaves are rich in fiber, potassium, iron, calcium, magnesium, phosphorus, vitamins A and C, the B vitamins thiamine and riboflavin, and protein. In *The Dandelion Celebration: a Guide to Unexpected Cuisine*, P.A. Gail reported that they are also nature's richest vegetable source of beta-carotene: they have 0.84 mg/g tissue; the carrot comes in at 0.61 mg/g tissue. Dandelions rank above broccoli and spinach in overall nutritional value. Ranked by nutrition, they're tied for 9th best, higher than romaine lettuce. Also, the roots are rich in iron, copper and other trace elements.

So, dandelions could be used to heal people suffering from all sorts of vitamin and nutrient deficiencies. Aside from that, they were used as a diuretic, which works because of the high potassium content of the plant. It's also the reason for one of its more irreverent common names, piss-a-bed.

Dandelion, as a food and medicine, was valued by many people of many cultures. And that's exactly why it has spread all over. People going to new countries prepared not just themselves, but their companion plants for the journey ahead. Dandelion was in the box of essential herb seeds, a hope for the future carried into new places.

Without a written record, there's actually no way of knowing the dandelion's precise origin story in the U.S., but current thought traces its arrival back to the 1600s: Michael Dillon, Emeritus Curator of the Chicago Field Museum of Natural History, tells us that when some of the nation's first European settlers started putting in their kitchen gardens, they planted dandelion for use as food and medicine. According to his research, it was brought to these shores on purpose, by the Pilgrims.

> "If it serves someone's fantasy, then yes, the first achenes [which are kind of like seeds, but not] were brought on the Mayflower in a small golden box, and with much ceremony and fanfare, Myles Standish planted them into the virgin soils of eastern North America,"

quipped Dillon, in an interview with Patty Wetli of Window to the World. But he had veracity behind his joke; the European colonists prized the plant for its many uses. The Puritans intentionally introduced it up north, the Spaniards down south, and the French planted it in Canada. Gardeners used to weed out the grass to make room for the dandelions. I'll admit a bias here: I think our ancestors were right about that. I have very little time for grass that isn't serving some purpose. Outside of playing fields, I think we should be planting more useful things!

Not too long ago, prize dandelions were exhibited at county fairs—one variety was patriotically christened the "American Improved."

Despite all that history, today an estimated 80 million pounds of pesticides are used to try and eradicate dandelions annually. How did it fall so far from grace?

There are a couple of threads to that tale. For one thing, dandelion was too good at acclimating to its new country. Far too good. Soon it was getting underfoot in all sorts of places it wasn't wanted, not to mention getting out into the native American ecology… which the Europeans weren't good at either noticing or caring about, if we're being honest with ourselves here. Still, it was making a nuisance of itself.

And then the lawn came into vogue. The English cultivated both the appearance and the concept of a lawn in the seventeenth century, playing off a Renaissance French ideal. In a time when most people were growing their own food, lawns were a symbol of wealth; a way to signal that a person could afford to maintain unproductive land. In this context, dandelions didn't symbolize spring: when they appeared in the perfect green turf, they were like scratches on a Lamborghini. Dandelions in the lawn symbolized neglect. Lax standards. Even—horror of horrors— poverty.

And so dandelions were relegated to the category of 'pest', and this little plant which had improved so many human lives, and should have been praised, became reviled.

If You Love It: You'd have good reason! And you'd be in good company. Much great art has been inspired by the resilient dandelion, including some stunning botanical works and the lyrical book *Dandelion Wine* by Ray Bradbury. If you're of a literary bent, I highly recommend looking up his meditations on the value of dandelions and their place in reminding us to stop and pay attention to small things.

Dandelions are a versatile little plant, full of good uses. The young leaves are healthy, and tasty in salads. The aerial parts are a nutritious cooked green, one of the first to raise its head in the spring. And it isn't only us that the plant can feed; the bees find dandelion nectar incredibly nutritious, and it helps them get through the bitter early spring.

The flowers make a wonderful wine. The tea made from the roots is effective as a diuretic and helpful for those looking for a coffee substitute… though it's not to everyone's tastes (I prefer the wine myself). Recipes including dandelion abound, and are easily found. I highly recommend experimenting with some of them!

If you want to gather wild dandelions, have at it! Just remember to pay attention to your area: if this is a place where herbicides might have been sprayed, don't forage from it. Also, a word of advice: avoid foraging by roadsides; heavy metals can build up in the plant tissues.

If you'd prefer to grow your own dandelions without the neighbors coming after you, I'd like to recommend French Dandelion. A wonderful supplier, Baker Creek Heirloom Seeds, has this to say about the cultivar:

> "A sophisticated French play on the immensely nutritious wild dandelion. Use the young leaves in salads, older leaves as boiled greens. Roots can be roasted and used in place of coffee, or lifted and forced during winter, like Belgian Endive. We're proud to offer this European strain of a traditional old favorite!"

I recommend buying seed from them, and they ship across the USA. Or, look into one of the decorative white and pink dandelion cultivars, perhaps? Dandelions are from the same family as dahlias, and with proper respect and care, they can be truly beautiful.

If You Loathe It: You're also not alone. And a few things have been figured out for dandelion control.

For the Home: It is impossible to completely control dandelion spread, given how their seeds are designed and how well they ride the wind. But it is possible to keep them from taking over your cultivated areas. Firstly, don't leave disturbed (aka cultivated) ground open. Nature does not like empty earth. It's not good for the soil. The way the natural world sees it, dandelions are like a bandage across wounded areas of empty ground. Nature, of course, doesn't understand that you just planted your tomatoes there. So, give nature a hand by mulching the soil. Anything organic can work for this in veggie beds: good quality straw, this season's fallen leaves (a great choice) or bark chippings. For established garden beds, I recommend a good quality wood mulch.

For Larger Areas: In lawns, a few tips:

- Never cut grass shorter than 2 inches and don't cut more than 1/3 of the leaf blade at a time. When you scalp your grass, you weaken the turf, set up the conditions for die-off, and let more light in for dandelions and other weeds.
- Cut down on fertilizer use, and use slow-release fertilizer. Since dandelions are expert at absorbing nutrients, they'll beat the grass to the fast-release stuff and get even bigger. Encourage plants like clovers to grow in the lawn, providing natural sources of nitrogen fertilizer.

In general, dandelion control is a matter of routine maintenance. Pull up rosettes with a weed-fork when you see them. Use a fire-wand or a pot of boiling water on dandelions coming up through cracks in sidewalks, rather than reaching for the herbicide. These are less damaging to the world in general, and about as effective in the short-term at keeping the place tidy.

The Chemical Option: If you're really struggling with a terrible outbreak, as in the case of, say, a long-neglected parking lot, you *can* go chemical. Go with 2,4-D or another systemic herbicide; top-kill products like Scythe are worse than useless on dandelions. The best time to use these systemics is in the fall, when the plants are transferring resources into their roots in preparation for winter. It is also possible to supplement this by putting down a pre-emergent product in the spring. But these are only for very specific situations, and aren't needed in regularly maintained properties.There's no pressing need to eradicate the plant for all time; you just need it to stop making your property look seedy... if you'll pardon the pun.

Dock

Rumex crispus, Rumex obtusifolius

Common Names: bitter dock, curly dock, curled dock, broad-leaved dock, bluntleaf dock, dock leaf, butter dock, yellow dock, patience, sour dock, narrowleaf dock, docken.

Family: Polygonaceae
Genus: *Rumex*

Looks Like: A tall perennial—and I do mean tall, from 2 to 5 feet tall!—the two closely related plants have erect, unbranched green stems with a reddish tint or streaks along the ribs. The large leaves at the base can be up to 14 inches long and 4 inches wide. They have blunt, heart-shaped leaves. One of the best identification features for docks is a small, thin sheath that covers the base of each dock leaf. This is called the ocrea, and it turns brown as the plant ages. Greenish flowers cluster tightly on their spikes at the very top of the plant. The small, one-seeded fruits are enclosed in a case with 3 toothed ribs.

Curly dock has wavy, curled margins to its leaves and its leaves tend to wind around the stem like a flag around a flagpole. Its flowers are whitish, fading into a dull brown. Bitter dock's leaves aren't as dramatic; they tend to be flat and broader than their curly cousin. Their flowers turn red as they mature, reaching a rich mahogany color in autumn. Curly dock is a biennial, while bitter dock is a perennial. Both plants roots can be as long as the aerial parts are tall.

I'm choosing to cover these two plants in one entry because they are practically synonymous in their uses and their history. They're both edible, and they're difficult to tell apart at a casual glance.

Comes From: Europe and Asia. At this point in the plant's spread, being more specific isn't going to happen.

Likes To Live: in disturbed areas with good sun; pastures, river banks and open places; stockyards, cowsheds and other places with high nitrogen levels. But it's not choosy; it's been found in the Arctic Circle. Dock is incredibly adaptable on water and soil; while it enjoys wet places, it's also a prolific weed of New Mexico. Annual rainfall? 14 inches. It's a stalwart of compacted sites. The only thing it doesn't like is acidic soil.

States Present: Every one of them. It's on the noxious weed lists for 46 states.

States Impacted: The states where animals are grazed or alfalfa is grown.

Legend
- List only
- Law only
- Both List and Law
- Included on other source
- Not Listed

Succeeds By: several mechanisms. Both species spread mainly by seeds, produced in very large numbers: up to 60,000 per plant, per season in the case of bitter dock. The seeds can be blown quite a ways on the wind, can float in water, and can survive a digestive tract. The teeth on the seeds also allow it to stick to fur, feathers or human clothing. Oh, and the seeds can stay buried for 21 years and germinate just fine afterwards.

Once it's established, bitter dock in particular has another trick up its sleeve. The roots branch after the first flowering and produce a set of spreading adventitious roots, which store up nutrient reserves over the next two years. This new root system sends up its own leaves and splits off from the original taproot after about five years.

Causes Trouble By: muscling out desirable plants and degrading aesthetic appeal of properties. It reduces production from grassland used to feed livestock, by reducing grass yields. Its feeding value for the grazing animal is only about 65% that of grasses. Digestibility of woody stems and inflorescences is only about 50% that of grass. Worse than that, it causes problems for the animals: in sheep, eating a lot of dock results in calcium deficiency, leading to sudden death or serious illness in the animals. It also shades out low-growing crops like alfalfa, causing a reduction in yield of the fields. Curly dock can also become a problem weed of no-till corn or soybean systems in domestic settings, the big weedy leaves and the ragged floral spikes detract from the appearance of a home and make the neighbors raise a brow.

Dock is one of those plants that has been a part of the Western world since the beginning. We get its Latin name, *Rumex*, from Pliny the Elder; this was his word for sorrel-type species. Hippocrates, Dioscorides, Culpeper, the Old English medical text known as *Bald's Leechbook* and the collection of Anglo-Saxon medical texts and prayers known as the *Lacnunga* all mention the dock plant.

A partial reconstruction of the flora of the central area of the city of Kiev, based on fossil spore and pollen samples, shows that several species of dock were already there in the 10th century AD. This makes sense if you know how many uses our ancestors put it to. It's part of the buckwheat family, and a number of sources mention turning the seeds into gruel. The leaves are a valuable spring and summer green. Under the name of patience, it was once grown in European gardens like a form of lettuce. Broad-leaf dock was used, because of its large leaf size, to wrap butter; hence one of its common names. It was used as a reliable mild laxative and a solid source of nutrients; both docks have approximately 4 times more vitamin A than carrots. The human body converts carotene into vitamin A, and this actually does increase night vision. Turns out your grandmother was right about eating your carrots!

All these nutrient and medicine-related uses are the most likely reason for the plant arriving in the New World. A study of traditional maize production in South America reported that bitter dock was already present in Mexico a scant few years after the arrival of the Europeans. The finding can be backed up by the plant's description in the *The Florentine Codex*, an ethnographic study of Mexico written by the Spanish Franciscan friar Bernardino de Sahagún around 1550. If the Spaniards brought it to the South, it's not too much of a jump to assume the French and the English brought it with them to the North. Pretty soon, bitter dock and curly dock put down roots anywhere their seeds could reach, and were making themselves at home.

If You Love It: You can put dock to all sorts of uses. The deep roots make a lovely golden-yellow dye when boiled. The leaves make a nutritious green, and both curly and bitter dock were widely eaten during the Great Depression in America. People appreciated the leaves for their tart, lemony flavor, their abundance, and of course for the fact that they were free.

Its flavor profile is on par with arugula, and it can be used in most of the same ways. Serve the greens with butter, bacon, and hard-boiled eggs. The leaves can also be stuffed like vine leaves, using a rice, herb, and cheese filling. Dried, they can be used as a seasoning for rice, potatoes, and seafood. Dock seeds can be boiled into a sort of porridge, or ground and added to flour for making bread, muffins and gravies. The stems of young plants can be chopped, simmered and sweetened with honey as a substitute for rhubarb pie. It's a good accompaniment for heavy meats, which it helps to brighten. It can be used as a substitute for lemon juice. The leaves are tastiest before flowering, but they're still very nice as a boiled green later in the season.

Do be somewhat careful: the entire plant contains a chemical called oxalic acid. In large amounts (if you ate dock leaves daily, for instance), it can cause kidney stones and a failure to properly process calcium. But if you treat it like spinach, you shouldn't have a problem. The seed heads are an important source of food for wildlife in winter.

Oh, and here's a tip from a girl born in a forest: if you're stung by nettles, find dock leaves, crush them in your hands and place them on the stings. Nettle is neutralized by dock. The old poem goes, 'Nettle out, dock in!'

If You Loathe It: You're definitely not alone. Many a gardener has faced a new plot of land full of dock plants with a sinking heart. But it can be controlled effectively.

For the Home: Your first tool is a fine-bladed shovel known as a sharpshooter. Sink the blade of the tool down beside the plant, and use it as a lever to work the deep root out of the soil. Don't forget, the roots will be as deep as the plant is tall! If the root breaks off in the ground it will grow back from the fragment, so watch out for new shoots from the pieces that managed to escape the blade.

Your next tool is fire. Burn the weed or throw it in the trash, never compost it. What you end up with is a compost pile absolutely covered in dock!

For Larger Areas: In large range situations, I'll share a quote from the Missouri State Extension:

> "Grazing and mowing can help reduce populations of curly dock. However, the weed has been shown to be toxic when consumed in large amounts. Curly dock seedlings can be controlled effectively with spring herbicide applications. However once the weed becomes established, fall applications may prove more effective."

The Chemical Option: For selective control of curly dock in grass pastures and hayfields, metsulfuron products (Cimarron, Cimarron Max, Chaparral, etc.), 2,4-D and dicamba combinations (Weedmaster, etc.), or combinations of GrazonNext or Grazon P+D with triclopyr (Remedy, PastureGard, etc.) are effective foliar sprays.

That said, if you're at home, stick with the shovel.

FoxTail Grass

Setaria viridis

Common Names: green bristlegrass, bottle grass, green foxtail, green millet, pigeon grass, wild millet.

Family: Poaceae
Genus: *Setaria*

Looks Like: a coarse, lime-green annual grass. The leaves are hairless and up to 12 inches long. Leaves unfurl from a curled bud. The leaf sheaths are hairless, except for short hairs along the margin. The mature plants are 1 to 3 feet tall, with erect stems that branch at the base. The flowers appear in June and persist through the growing season; cylindrical, with crowded spikelets that each have 6 to 10 long, greenish or yellowish bristles that stick like velcro. Seeds are oval and greenish to dark brown.

Comes From: Northern Africa.

Likes To Live: in cultivated fields, gardens, waste places, disturbed areas and roadsides.

States Present: All fifty.

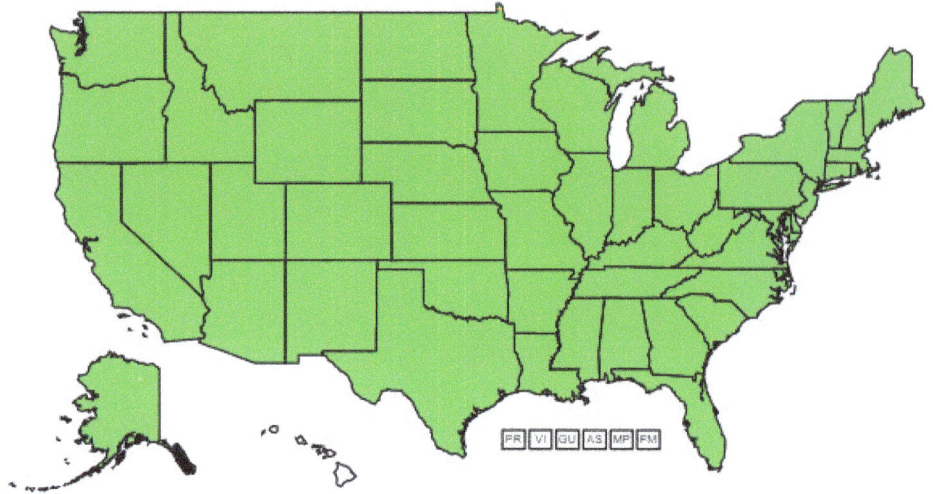

States Impacted: All the farming states.

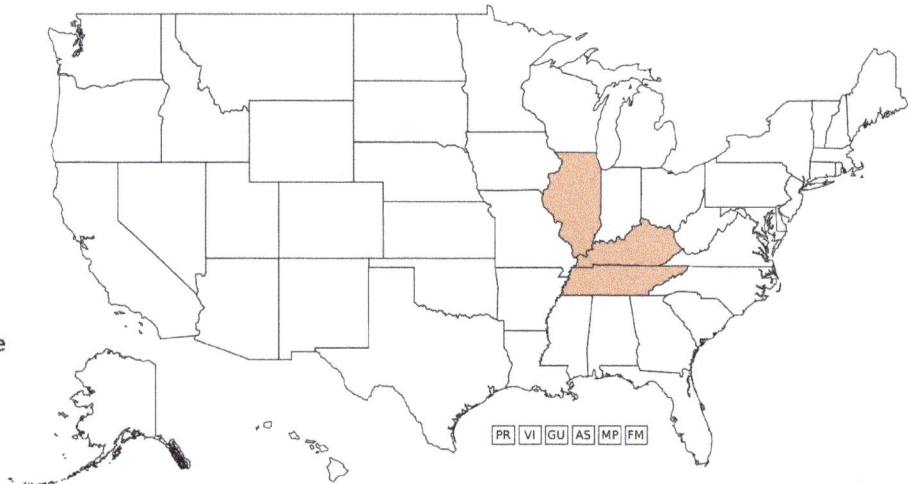

Legend
- List only
- Law only
- Both List and Law
- Included on other source
- Not Listed

82

Succeeds By: seeding prolifically and spreading those seeds everywhere. They stick to fur, hair, clothes, and the seed heads (awns) can burrow into skin. It is not nice. At all.

Causes Trouble By: Foxtail grass in a field can easily outcompete cultivated crops and grains, reducing the yield by up to 36% in maize, sugarbeet, soybean and cereal crops. It can damage the price at which a farmer can sell their crop seed when it's present as a contaminant. In addition, it regularly does harm to domestic animals by causing skin infections, eye damage, or by causing breathing problems when the seeds are inhaled.

"Look out, it's foxtail season." You hear it a lot around the Western states. "Look out for the foxtails." Why? Because if you have pets, you must be vigilant of the weed. It can irritate, wound, or even kill your dog or cat. Foxtail awns—the barbed, razor-sharp needle tips on each seed designed to drill the seeds into the ground—can embed themselves in skin. An animal body can't break down the tough plant material of foxtail awns, so it can lie under the skin and cause a localized infection.

When it migrates, its barbs keep it moving on a one-way journey through the body. The foxtail awn, carrying bacteria with it, keeps tunneling into tissue, carving the dangerous path of infection that marks grass awn disease, more commonly known by the dread name of ripgut. Organs can be pierced, fungal infection can arise, and bacteria delivered deep inside the body packs a nasty punch.

Breathing can draw awns further into nasal passages. Inhaled foxtails can travel from the nasal cavity to the lungs; a common site for deep-seated infection. If it reaches the peritoneum, the inner lining of the body, or the deep tissues, it will cause internal abscesses that can kill when they rupture.

Foxtails often end up embedded in the mouth. From there, they can work their way into the deeper tissues of the head and neck. When they do, animals present with a mysterious deep abscess around the throat area that has to be drained and cleaned by a veterinarian. Warning signs of foxtail awns in your pet include extreme sneezing, head-shaking, coughing, licking of a skin puncture, and—most worrying of all— a high temperature. It's a terrible thing, ripgut. It's heartbreaking when it happens to a domestic pet. It can ruin livestock farmers when it kills their animals.

So why was such a beastly plant ever brought to America? It wasn't brought here, in fact. It came here all on its own. In North America, foxtail grass is first reported in Montreal, circa 1821. Genetic studies and historical records pinpoint this time and place. The hypothesis is that the seeds were introduced unwittingly, as a contaminant in bags of crop seed. It also arrived as a contaminant in the ballast of ships, which was traditionally stones or sand, picked up before departure and dumped upon arrival. Oops.

Foxtail grass wasn't too bad as a weed in Canada; the cold kept it under control. But a couple warm winters in the 1930s let it really get going, and by 1948, the species took off throughout Manitoba, Alberta and Saskatchewan. In the USA, foxtail grass has been present since at least 1900, and as the climate has changed in the last hundred years, it has thrived.

If You Love It: I'd be much surprised. It is edible, but only in the 'won't kill you' sense of the word. There are much, much better things to eat. I've found no records of any particular uses.

If You Loathe It: Then let's talk control. There's several approaches.

For The Home
- **Weeding:** Persistent weeding can get this plant under control. Dig it up, double bag it and throw it out.
- **Fire:** Burn the plants, and they die. With the obvious caveats about safety, this is an effective non-chemical approach.
- **Soil Improvement:** Foxtail grass competes particularly well in degraded soils. Improve the soils by adding nutrients, organic material, and aerating the soil, and you should have much better results.
- **Mulch:** Mulching the area can stop the plant's seeds from growing

For Larger Areas:

- **Goats:** They aren't affected by foxtail, and will happily eat it. Look into a service that rents out goats for organic lawn care.
- **Tilling:** Foxtail will die if it's pulled or tilled, so use a hoe in small spaces or a tiller in large spaces, and yank it out of the ground.

The Chemical Option:

Since the plant is an annual, a good pre-emergent herbicide granule spread evenly will prevent the seeds from sprouting. Put it down in late February (depending on your zone) for three seasons, and you won't have any more foxtail by the fourth spring.

Giant Hogweed

Heracleum mantegazzianum

Common Names: cartwheel-flower, giant cow parsley, giant cow parsnip, wild mustard, white mustard, hogsbane.

Family: Apiaceae
Genus: *Heracleum*

Looks Like: a biennial that can grow up to 15 to 20 feet in height. The stem is hollow, and usually blotched with purple. Both the leaf stalks and stem produce bristles. The stem can be 2 to 4 inches in diameter. Leaves are lobed in many places, giving them a lacey, ruffled look. They can be up to 9 feet wide. Flowers bloom from June to July. The inflorescence has many white florets that form a flat-topped umbel. Each inflorescence can have a diameter of up to 3 feet. They dry into fruits that have brown resin canals which can be up to a 1/4 inch in diameter.

Comes From: the Caucasus Mountain region of Central Asia.

Likes To Live: along streams and rivers and in fields, forests, yards and roadsides. It likes open sites with good and moist soil, but it can grow in partly-shaded conditions too.

States Present: Maine, Vermont, New York, Massachusetts, Connecticut, Pennsylvania, North Carolina, Michigan, Indiana, Illinois, Wisconsin, Michigan, New Mexico, Oregon, Washington, Alaska.

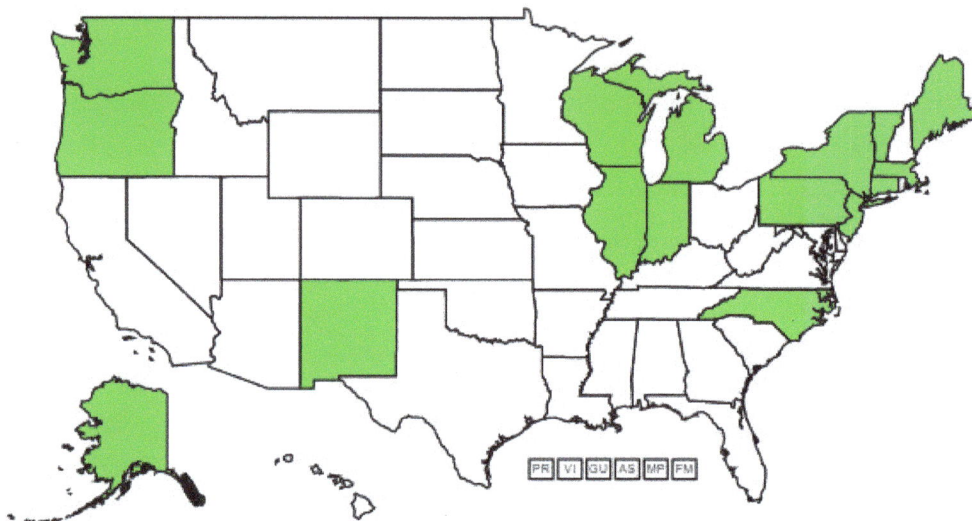

States Impacted: Everywhere it grows.

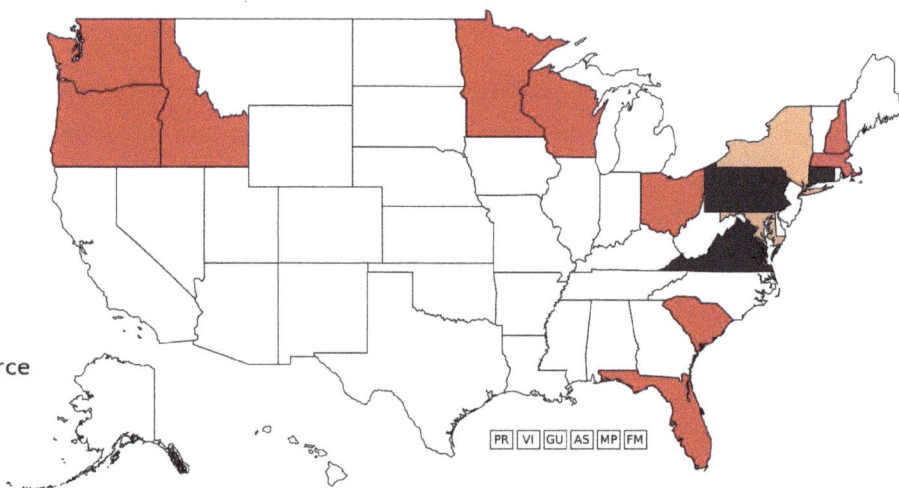

Legend
- List only
- Law only
- Both List and Law
- Included on other source
- Not Listed

Succeeds By: seeding incredibly—from 5,000 to 100,000 per plant, per season—leafing out earlier than most plants to get a competitive edge on the native flora, being self-fertile and being absolutely immune to all predation. Nothing touches giant hogweed if it knows what's good for it. Its seeds spread by water, where they can float for up to three days, and by wind.

Causes Trouble By: being a big, miserable bully. When it establishes, it goes about taking up all the available room, destroying species diversity and poisoning everything that gets near it. It's a champion at outcompeting species for habitat, especially in riparian zones, and causing soil erosion. The enormous height and leaf area of tall hogweed enable the plants to loom over most indigenous plant species and outcompete them for light. In established stands of hogweed, up to 80% of the incoming light is absorbed by the plant. Giant hogweed craters the species diversity of the native plant communities. Enough giant hogweed seeds getting into the ground can absolutely destroy the native ecology of an area it likes. It tends to grow into waterways and ditches, and eventually it can block the flow of water. And that isn't all. If anyone tries to move the problem plants, they get a very nasty surprise. This plant is on the federal noxious weed list because of its poisonous sap. The latex makes mammalian skin very sensitive to UV radiation, causing large blisters and severe chemical burns.Large doses of furanocoumarins, the responsible chemicals, can cause cancer or fetus malformation. The damage the sap can do makes giant hogweed very dangerous to work with. The plant is a danger to the public, and eradication efforts are costly to the impacted states.

This is one of the invasive plants that makes me cover my eyes with one hand and groan. The first record of introduction from its native range was a specimen planting in 1817, when the Kew Botanic Gardens of London recorded it in their seed list as a 'botanical oddity'. In 1828, the first wild population was recorded in Cambridgeshire, England. Already, the plant had begun to spread. And it didn't stop there. Of the 19 countries we have records for, 14 countries describe the plant's arrival before 1900. Two have first records between 1900 and 1960, and everybody else had it by 1960.

And here's what really makes me groan. It didn't just escape. It was introduced as a decorative ornamental! Seeds were traded, bought, sold, and planted in botanic gardens and the grounds of well-to-do estates. It was also a favorite of beekeepers, due to the size of its flower heads. They weren't completely wrong; the amount of food for bees is substantial. At least the things were good for *something.*

This fashion for planting giant hogweed as a dramatic centerpiece or backdrop continued for most of the 19th century, an incredibly long time for everyone to miss the glaring problems with the species. You would think that a plant that will literally take your skin off if you brush it would turn people off fairly quickly, but oh no. It was *fashionable.* Granted, these were the same people who used arsenic to color their own clothing, so I suppose it's no surprise.

The fad finally died out after warnings about its dangers finally started showing up in Western European literature towards the end of the 1900s. Unfortunately, by then giant hogweed was planted in the gardens of Highland Park in 1917, in the city of Rochester, New York. It had been introduced to American upper class gardens as a 'botanical curiosity'. And we were stuck with giant hogweed ever after.

If You Love It: Consider your life choices, perhaps?

If You Loathe It: Grab your protective clothing and your gauntlets. It's time to talk about eradication.

For The Home: Do *not* move soil contaminated with seed. If removal of contaminated soil is necessary, hogweed seed sprouting can be prevented by burying the soil to a minimum depth of 20 inches, and covering it with clean soil.

Since the dispersal of giant hogweed is almost entirely by seeds, it is very important to prevent the plant from setting seed. Control new, smaller infestations first, before the seed bank has a chance to establish.

Wear safety clothing when dealing with giant hogweed. **All skin must be covered!** Wear:
- Pesticide gauntlets with a pair of cotton gloves beneath
- Face guard
- Goggles
- Hat
- Heavy, long sleeved shirt
- Jeans
- Work boots

Manual & Mechanical Control: All methods will need to occur for multiple years until no new plants grow from the seed bank. Monitor the site for at least 3 more years to make sure no new seedlings appear. The methods are:

- **Root Cutting:** This method is ideal for a single plant or small infestations. Cut the taproot at least six inches below ground level using a spade, shovel, or anything with a large, sharp edge. In areas with erosion or on steep inclines, plants may need to be cut up to ten inches below the soil. When cutting the root, you need to separate the thick stem base bearing old leaf scars from the root below. If you can do so safely, cut the plants before the leaves are very large. The cut part of the plants should be removed from the soil and left out to dry, or double-bagged and disposed of.

   Do Not Burn This Plant. The dangerous chemicals in the tissues will volatize, and if you breathe them in your lungs will blister. If the plants do not die, cut them again when they regrow. When using the root-cutting method, another attack strategy is to apply herbicide to the cut root remaining in the soil to help ensure no part of the root grows back the following year.

- **Hogweed Flower Head / Seed Head Removal & Safe Disposal:** Cut off the flower heads after the seeds have formed, but before they mature. This will prevent the plants from shedding the seeds and forming new flowers on lower branches. Use a pruner, lopper or knife to carefully cut off all flower heads and place them in a sturdy (or doubled) trash bag. Clear trash bags are the best for solarization (a method of using the heat from the sun to destroy the seed viability) but dark bags are suitable as well. Seal the trash bags tightly. If the sap has gotten on the outside of the bag, put it inside another trash bag so the outside is safe to handle.

> Place the bags in a secure location where they will be exposed to direct sunlight to give flowers and seeds the solarization treatment for at least two weeks. Then, dispose of the trash bags in the garbage. It is important to solarize; otherwise, you're simply encouraging the plant to start growing at the city dump.

- **Cut & Cover:** Cut the plants down to ground level and cover the soil with black plastic or landscape fabric with mulch on top. Check the following year to make sure seedlings do not poke through. After a few years, remove the plastic and re-vegetate the area.

- **Monitor:** Hogweed seeds can be viable in the soil for 15 years. Long-term monitoring is absolutely essential. Check the site and surrounding area for the next several years before you relax and declare yourself the victor.

- **Revegetation:** This step is crucial! After removing giant hogweed plants, you'll be left with an area of bare soil, vulnerable to soil erosion, a return of the hogweed, or some other invasive weeds. Mulch the site and plant it with something good.

For Larger Areas:

- **Cutting & Mowing:** Start when plants are small and continue mowing throughout the season. DO NOT mow if the plants are larger than your mower and NEVER mow if there is a flower or seed head. If you choose this method, the plants must be mowed at least three times during a growing season, for several years. Repeated mowing is often used for large infested areas and, if done consistently, can destroy most of the plants.

Be careful not to spread the seeds. All flower heads should be removed and carefully disposed of prior to mowing. Mowing equipment should be cleaned before using in another area, to avoid spreading hogweed seeds. Remember that the sap is toxic, and wear protective gear!

- **Monitor:** Hogweed seeds can be viable in the soil for 15 years. Long-term monitoring is absolutely essential. Check the site and surrounding area for the next several years before you relax.

- **Revegetation:** Seed the site with grasses and/or natives. Only then will it truly be saved from The Giant Hogweed Of Doom.

The Chemical Option: Giant hogweed is vulnerable to systemic herbicides, such as glyphosate and triclopyr. These herbicides can be used for the control of a single plant, or large stands of giant hogweed. These systemic herbicides will be absorbed by the leaves and will move into the root, killing the plant completely. Triclopyr is a selective herbicide that acts only on broadleaf plants and will not harm grasses in the area. Glyphosate is a non-selective herbicide, and will kill any plant it touches. Some triclopyr and glyphosate products (for example, Rodeo) are labeled for use in riparian areas and near water. Just read your labels very carefully! If you're nervous, contact your local extension. Areas sprayed with triclopyr can recolonize with grasses and other herbaceous species within the same growing season, which helps to suppress giant hogweed growth and cuts down on the soil erosion. Recolonization at sites using glyphosate will be slower than sites using triclopyr.

If you decide to spray, use the recommended manufacturer's dose and follow label instructions very carefully! Apply the herbicide between March and early June, when hogweed leaves are actively growing. A follow-up treatment in July or August may be needed for the plants that survive the first round. During this follow-up treatment, remove any flower heads present to cut down on the seed load. I personally favor the option of carefully cutting the plants at the basal rosette (the bottom leaves) and spraying those, disposing of the rest. Any control method will have to be repeated for a few years. But the beast can be defeated.

If you get sap on your skin:
- Wash your skin thoroughly with soap and water as soon as possible.
- Protect your exposed skin from sunlight for at least 48 hours.
- If you experience a skin reaction, call your healthcare provider.

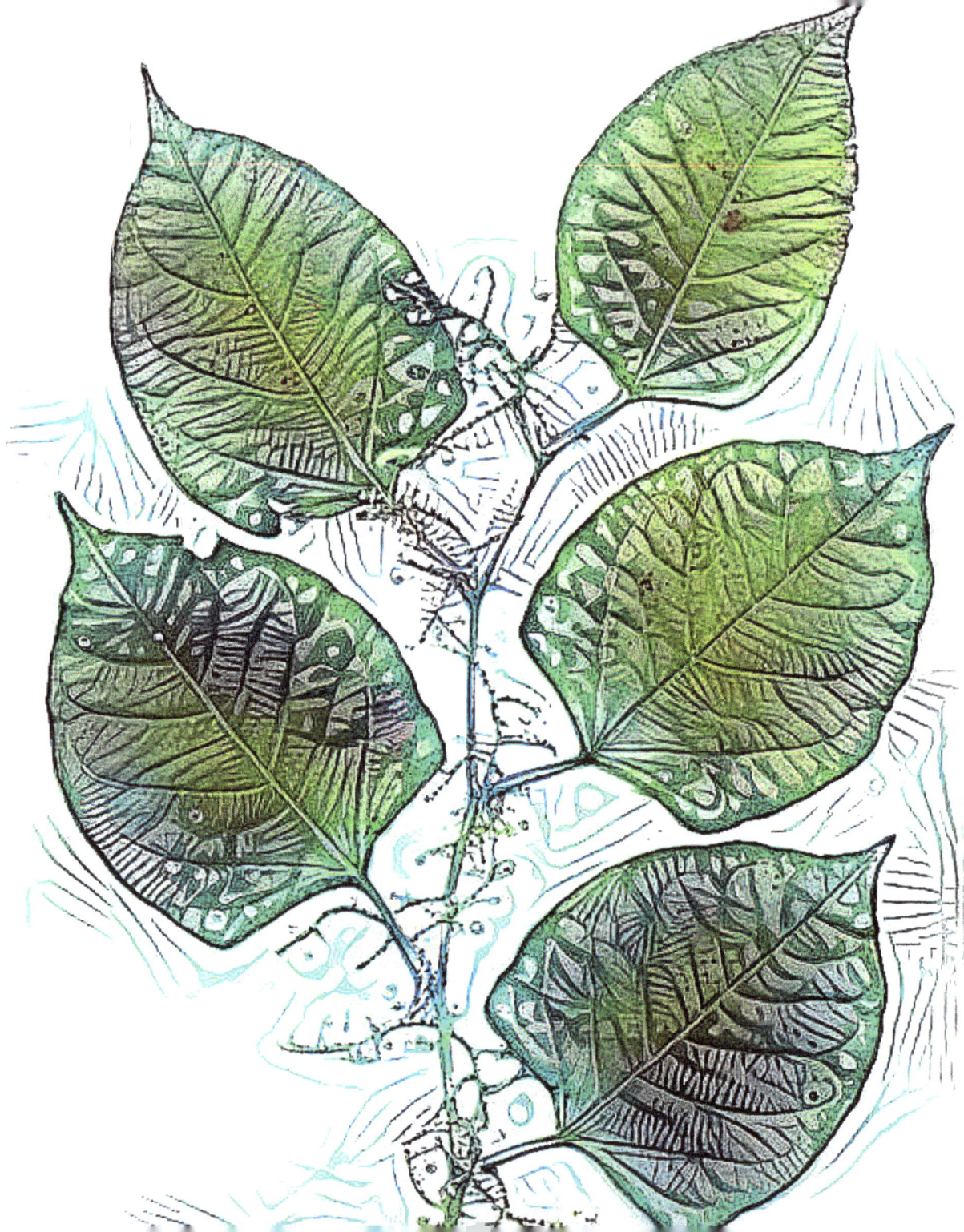

Japanese Knotweed

Reynoutria japonica

Common Names: Huzhang, Itadori, Asian knotweed, Bohemian knotweed, donkey rhubarb, gypsy rhubarb, Sally rhubarb, German sausage, Hancock's curse, Devil's bamboo, pea-shooter plant, elephant-ear bamboo, fleece flower, Japanese bamboo, Japanese fleece flower, Mexican bamboo, Devil's weed, demon's weed, demon weed.

Family: Polygonaceae
Genus: *Reynoutria*

Looks Like: an attractive deciduous shrub, reaching heights of 10 to 16 feet. The semi-woody stem is hollow, with enlarged nodes. Alternate heart-shaped leaves are 6 inches long and 4 inches wide. Flowering occurs in late summer, when pretty tassel-like white flowers develop in long panicles along the axils of the leaves. The plants are dioecious (male and female flowers occur on separate plants). The fruit are papery and winged, containing shiny, black, three-sided achenes.

Comes From: Eastern Asia. In Japan, the plant is not nearly as problematic as it is in its introduced range. However, since its native place in the Japanese ecosystem is to act as a colonizing species on recently erupted flows of volcanic ash and lava, it laughs at asphalt. You can imagine how much fun it has with something as easy to work with as loam.

Likes To Live: in disturbed areas with high light, such as roadsides and stream banks. It thrives in disturbed sites, such as roadsides and construction areas. It will invade lawns, home properties, and anywhere with sun. It can grow in shade, and in all sorts of soil conditions. Powerful roots can work their way through cracks in concrete and deep into pipes, seeking nutrients and water.

States Present: So far, only Florida, Nevada, Arizona and North Dakota haven't been cursed with it. Everybody else has it. It's on the Noxious Invasive Weed list in 36 states.

States Impacted: Every state where it grows. The species is widely considered to be one of the worst invasive plants in the temperate world, and its propagation and distribution is banned both in America and most of Europe. It costs the United Kingdom about 150 million pounds a year to control the plant, and in the USA the cost is 300 million dollars.

Legend
- List only
- Law only
- Both List and Law
- Included on other source
- Not Listed

Succeeds By: spreading through seed dispersal, independent clonal propagation through small vegetative fragments, and by rhizomes, Even the smallest root and stem fragments can resprout with a vengeance, making this plant extremely hard to eradicate.

Causes Trouble By: terrorizing every locale it gets its rhizomes into. In the spring the new stems elongate incredibly fast, and in a matter of weeks they have put out a green canopy that blocks out most of the light. Every time a bit of branch or stem is knocked off or cut, it falls to the ground or into any available water, and is spread to some new place to sprout—which it does, with a terrible enthusiasm. The seeds have some impressive sprouting ability as well, but it's the incredible abilities of the plant to propagate vegetatively that makes it so dangerous. Its rhizomes spread it out in all directions as well.

And what does it do when it spreads? Oh, a few things. First, it blocks out the light below it completely. And since it can grow up to 16 feet high, everything is beneath it. Anything trying to grow beneath japanese knotweed will wither away for lack of sun. This drives away the native fauna, who no longer have anything to eat. Soon, the knotweed is the only plant in the area, absolutely decimating and destroying the native ecology. It can also make the riverbanks it grows on erode more quickly, and the thick stands of knotweed can block waterways and small rivers, causing flooding.

And that's not all. Not only does it eat your garden, it eats your home's property value. In the United Kingdom, sellers have to disclose the presence of Japanese knotweed on deeds. Banks require a written management plan if it is present on or near property. They will not underwrite a mortgage without one. Essentially, the plant can make your property worthless.

Why? Because if you can see Japanese knotweed, the house is already in mortal danger. Japanese knotweed can grow up through the tiniest crack, taking advantage of any water that can go down the seemingly harmless crevices in the first place. Structural fixtures and fittings are particularly susceptible to the colonization of Japanese knotweed, with just the smallest hole on a section of pipe or joint being more than enough to encourage the knotweed's water-seeking rhizome. This rhizome, once tapped into the source, will continue to grow and eventually block the hole, in many cases eventually cracking and breaking it. Buildings are equally vulnerable. The plant can and will grow up inside the spaces between walls, searching for nutrients and warmth. Once inside, it has the strength to expand the internal space of the wall until it cracks the casing. It can work its way up through floorboards, crack brickwork, and fill up utility cable casings with itself. Japanese knotweed can also dig its rhizomes into concrete-slab foundations and destroy them slowly but surely. If you see the plant, the damage has probably already begun. Consider the two quotes below:

> "This knotweed is one of our most important introductions from Japan, a perennial ornamental plant, inextirpable, with shining foliage, clusters of very graceful flowers, useful in creating groves, sheltering young plantings, and fortifying sandy hill sand dunes. The plant, which can be cut in the spring many times over, provides an excellent forage for fattening livestock, which eat it out of preference; the flowers, which appear in autumn, are very sweet and give bees winter food; the bitter and tonic root is a medicine of repute among the Chinese and Japanese; finally, even the stalks which die in winter are good for burning and for matches. Already there have been very satisfactory trials stabilizing trenches and slopes along railroad tracks and sandbanks with plantings."
> —Von Siebold & Company Nursery, United Kingdom 1848

> "Under section 14(2) of this Act, it is a criminal offence to plant or otherwise cause to grow in the wild any plant which is included in Schedule 9 of the Act including Japanese knotweed. The maximum sentence for the offence on indictment is an unlimited fine and/or two years' imprisonment."
> —Wildlife and Countryside Act, United Kingdom 1981

In the course of a hundred years, both these things were said about the same plant. And herein lies a tale.

The worldwide infestation of Japanese knotweed can be credited to a single man: Philipp Franz von Siebold, Surgeon Major for the Dutch at Nagasaki between 1823 and 1829. An avid amateur botanist and a man of both brilliance and (possibly a bit amoral) ambition, Siebold helped to found the Royal Society for the Advancement of Horticulture in the Netherlands. Siebold circumvented Japanese regulation on plant exports, dodged their attempts to keep foreigners from learning too much about their world, and shipped 500 species of plants back home from Japan over his time there. Many of the plants he introduced, or their cultivated descendants, are such staples of American gardens that we think of them as American plants: Siebold forsythia (Forsythia suspensa) and PeeGee hydrangea (Hydrangea paniculata 'Grandiflora'), for example. His final shipment was sent in 1842. In it was a graceful bamboo look-alike that Siebold hoped to coax into surviving colder northern winters. He started a nursery to monetize his interesting Asiatic finds, dubbed it Von Siebold & Company, and established his planting operation in Leiderdorp, just across the river from Leiden, Holland.

In his 1848 catalog, von Siebold first offered Japanese knotweed for sale to the public (under the name Polygonum sieboldii) at the exorbitant price of 500 francs for a "mother" plant, along with 25 young ones. In 1850, Siebold brought a specimen of Japanese knotweed to the Utrecht plant fair in the Netherlands. He enthused on its virtues, describing it as 'growing vigorously' It won the gold medal at the fair, and was prized for its 'gracious flowers.' It also won him a higher profile for both his nursery and his interesting plant.

He continued his marketing by sending samples of Japanese knotweed to Kew Gardens and Royal Botanic Gardens in Edinburgh in the 1850s. And it sold. Oh my, how it sold. It was selling so well that Siebold increased production. Eight years later, in his 1856 catalogue, the price had dropped to 100 plants for 25 francs. By 1869, it was widely available for sale to the public, advertised in glowing terms by nurseries, and even being planted by farmers for animal feed.

At some point around 1873, the plant arrived in America. Folk history tells us that Frederick Law Olmsted, famed American landscape architect, is responsible for introducing Japanese knotweed into the United States. Histories of New York and guides to noxious weeds tell us that the great man planted it in Central Park, and along the Muddy River in Boston's Emerald Necklace. We do have documented proof that Olmsted met William Robinson when Robinson visited the United States around 1870, and that he later owned a copy of *The Wild Garden*, which he recommended to his partner Calvert Vaux during the laying out of The Ramble in

Central Park. The book repeatedly encourages the planting of Japanese knotweed, so the idea isn't impossible. But since the plant isn't on the original ordering lists for either the Muddy River or The Ramble, that bit of lore might or might not be true. Ann Townsend makes some very credible points against the little bit of lore in her work, 'Japanese Knotweed: A Reputation Lost', and I'm inclined to agree with her in doubting it.

For decades, how the weed really got to the USA has been a puzzle. But Peter Del Tredici may have finally found us the answer. I'll quote from his paper on the subject:

> "In the course of doing research on the plant introductions of Thomas Hogg, a famous New York City plantsman who sent numerous plants from Japan to America between 1862 and 1875, the author has discovered the earliest known reports of the cultivation of Japanese knotweed in North America. The first article appeared in the March 1868 issue of American Agriculturist under the title, "The Japanese Knotweed (Polygonum cuspidatum)." The article, which is accompanied by a line drawing of a plant that is clearly producing young fruits, was written by the journal's editor, the botanist George Thurber. After giving a general description of the genus Polygonum, Thurber (1868) goes on to provide the following description: We have for some years known a species which is really worth cultivating, but which does not seem to be much disseminated—the Polygonum cuspidatum, a native of Japan. It is a perfectly hardy perennial, which throws up a branching stem three or four feet high, bearing large oval leaves, which are long pointed at the apex,—hence the name cuspidatum. The small white flowers are succeeded by the fruit, or seeds, which being of a pale rose color are more showy than the flowers themselves. Though the flowers individually are small, they are produced in such abundance and have such a graceful drop that the plant is quite showy in flower and fruit, and its effect is heightened by the reddish color of the stems. The plant increases rapidly, and soon forms a large clump, indeed this is its greatest fault, and one which makes it unfit for use in small borders. It is very effective for planting where there is plenty of room, and it will grow in any soil and situation, even under the shade of trees. It blooms in July and August, and continues for a long time."

It seems that there was quite a love affair with this plant on both sides of the pond. But slowly, very slowly, the romance began to wear off. By 1887, gardeners in the parks of Oldham, England, were lamenting that knotweed "kept appearing in nearly every piece of cultivated ground." In 1899, the celebrated British gardener Gertrude Jekyll urged that knotweed should be "planted with caution," and by 1905, the UK's Royal Horticultural Society was anxiously stating, "it must be kept in check."

Too late, the warning came. Far, far too late. The plant had spread far and wide in the UK. In the USA, Japanese knotweed has colonized the Northeastern United States, the spine of the Appalachians, the Great Lakes states, and the Pacific Northwest.

DNA analysis of 150 knotweed samples from across the UK showed them to be identical, meaning they were all clones of that medal-winning plant Siebold brought to the Netherlands, 150 years before. Siebold had been right in one thing: it did, indeed, prove to be 'inextirpable' and 'grow vigorously.' As it still is to this day.

If You Love It: I'd be surprised, but the plant isn't completely useless. Knotweed tastes similar to a lemony rhubarb, and eating it is certainly one way to dispose of the plant material. In Japan, knotweed is known by the name Itadori, meaning 'heals the sick' or 'heals pain'. The volatile oils emodin and phsycion, present in the roots and leaves, do demonstrate anti-inflammatory effects. And other active chemicals, including resveratrol, resveratroloside, polydatin, and piceatannol, show provable antioxidant activity. It probably is good at reducing inflammation and enhancing wound healing when used as an external poultice or turned into a cream. And it is good as a green. The resveratrol, in particular, makes it a healthy addition to the diet. The good old *Peterson Field Guide to Edible Wild Plants*, a favorite book on my shelf, recommends steaming or boiling the stems for about 5 minutes and treating them either like asparagus or rhubarb.

If You Loathe It: You're definitely not the only one. Controlling knotweed is a long and difficult process that may well be best to turn over to an experienced contractor. But I'll outline the basic steps and issues here.

For The Home

- **Mapping:** Walk the property and get a sense of the scale of the infestation. Take into account the location and extent of Japanese Knotweed, as well as a planned buffer zone of at least 6 feet from all structures including fences and all underground features, such as piping, sumps, and wiring.
- **Control:** Now, here's the tricky bit. Killing the stuff. For small infestations, dig or pull the plant by hand or with shovels. *Don't* throw these plants in the compost heap or the local trash; burn them. Otherwise, you will be responsible for spreading the infestation. Don't mow Japanese knotweed as a control method; mowing will spread fragments of the plant to new areas, where it definitely will take root.

> Take all possible measures to avoid spreading even a half-inch of the plant material to a new site. If you can't burn the plant, contact your local county extension officer for disposal options.

> Digging up the contaminated soil to a depth of 2 feet, replacing it with clean fill dirt and disposing of the infested soil is an effective way of removing Japanese knotweed roots. Be aware that the contaminated soil is considered hazardous waste. Contact your local yard waste or compost site to see if they accept noxious weeds, and get your site checked for utilities before you sink the shovel.
> You can only transport noxious weeds to a disposal site. Dumping it in any general landfill is illegal, and the fines are high.

> Watch out! Like the undead, Japanese knotweed can rise from the dead after three years of dormancy. Be vigilant for at least five years.

- **Restoration:** Restoring the infested land is a critical aspect of invasive plant management. Site restoration will result in a healthier ecosystem, which will help to fight future invasions. Plant the area with native plants, or mulch and plant healthy garden plants in a garden setting.

- **Invasive species reporting:** Populations of Japanese knotweed and other invasive species can be reported directly into the free mapping database program EDDMapS (www.eddmaps.org). This program maps the locations of known invasive species on a nationwide scale. It can be used to determine potential problem areas, and help track newly detected invasive species outbreaks for Early Detection & Rapid Response measures. EDDMapS can be accessed via their website, or by using the Outsmart Invasives app. Once the report is submitted, it goes to the state verification group, gets approved, and posted. The information is then available to the public. It can be a valuable resource for anyone involved in invasive weed control, natural area remediation, or construction.

The Chemical Option: Combining mechanical and chemical approaches is a good idea when you're up against knotweed. Using a systemic herbicide like glyphosate (Roundup), follow best practices. Spray on a calm, cool day. Apply the spray directly to the leaves; you'll need 6 to 12 inches of healthy green tissue to take the poison up properly. Spray the plant once a month for a season. Yes, it really does need that much.

Jimsonweed

Datura stramonium

Common Names: Jamestown weed, mad apple, moonflower, stinkwort, thorn apple, devil's trumpet, devil's snare.

Family: Solanaceae
Genus: *Datura*

Looks Like: a large, annual herb which grows up to 5 feet tall. It has pale green stems with spreading branches. Leaves are ovate, sometimes with a purplish tint, coarsely serrated along edges and 3 to 8 inches long. The plant blooms from midsummer to first frost. Flowers are creamy white shading to purple, with a 5-pointed corolla arranged in a pinwheel folding form, up to 4 inches long and set on short stalks in the axils of the branches. The seeds are contained in a hard, spiny capsule the size of a golf ball, which splits into four segments when ripe. Overall, it has an appearance that's kindly described as 'exotic', or less kindly described as 'witchy'.

The flowers have a lovely scent. The rest of the plant is much less pleasant; it smells like unwashed feet when you crush the leaves. **Beware:** the plant contains alkaloids you don't want on your skin. Wear gloves. The numerous seeds are dark brown to black, flat, kidney-shaped with an irregularly pitted surface, around 1 to 1-1/2 inches.

Comes From: South America.

Likes To Live: along roadsides, railways, disturbed land, wasteland, fallow land, crop fields, managed pasture, drainage ditches, woodland edges/gaps, lowlands, gullies and dry riverbeds.

States Present: Everywhere but Idaho and Wyoming… for now.

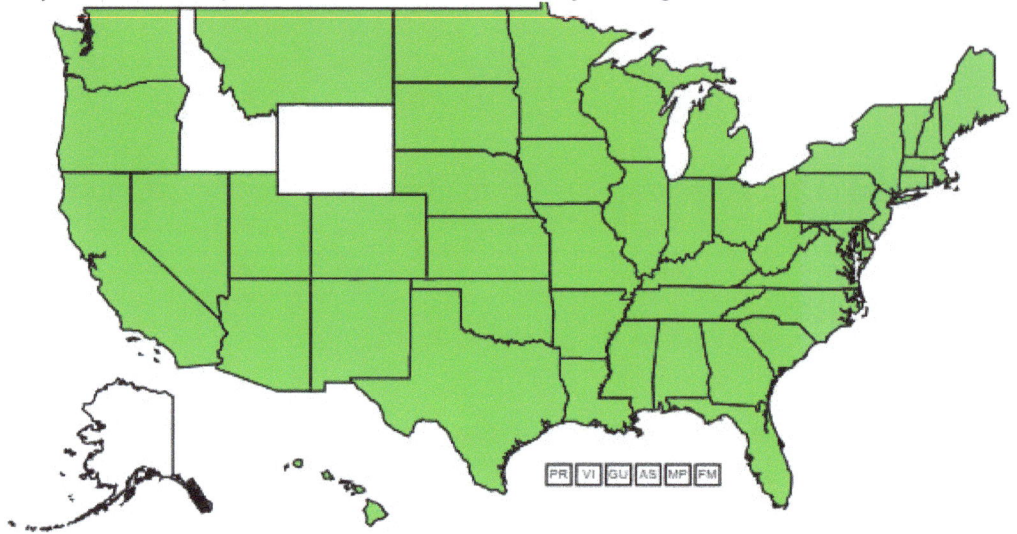

States Impacted: Anywhere raising animals or growing cotton, soybeans, peppers, potatoes,

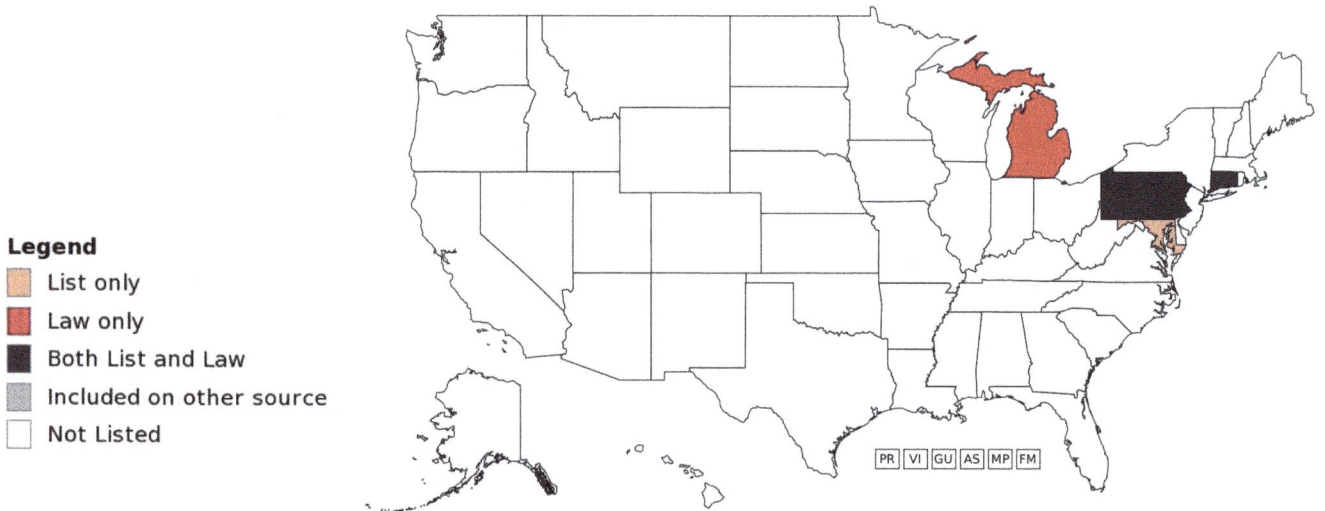

Legend

- ▢ List only
- 🟥 Law only
- ⬛ Both List and Law
- ▨ Included on other source
- ▢ Not Listed

Succeeds By: SEEDS. An annual plant, jimsonweed reproduces by seed, and it can produce a *lot* of seeds: up to 30,000 seeds per plant! And the seeds are viable for up to 50 years. The plant's also self-fertile, so a single plant can kick off an infestation.

Causes Trouble By: competing aggressively with native plants and crops, forming dense stands that shade out everything else. Infestations in the USA cause a 56% yield loss in cotton. Yields of soybean plants growing near jimson weed are significantly reduced, showing the allelopathic impacts of this weed. In Spain, competition from thorn apple in irrigated maize has reduced yields by 56%. Since it is extremely poisonous, it absolutely must be cleaned out of all crops, at some expense to the farmer.

It's also an alternative host for several pests and pathogens in the solanaceae crops (potatoes and tomatoes). Toxic to people, horses, cattle, sheep, pigs and chickens, it can drop the value of land it grows on. It also has a social cost: children are poisoned by it every year. It has gained a nasty notoriety as a plant that is used by people committing suicide.

Here's how the old story goes, more or less: 1676, Virginia. A few British soldiers were camped outside Jamestown. Some versions say they were there to deal with Bacon's Rebellion. Some don't. The soldiers were cold, disheartened, and starving. Looking around their campsite, they popped whatever plant matter they found in an old tin pot and boiled it up for a sort of soup (or, in some versions, a salad). That night, two of the soldiers died. Five went mad. One went blind. Robert Beverley Jr. wrote about their delirium in *The History and Present State of Virginia, Book II: Of the Natural Product and Conveniencies in Its Unimprov'd State, Before the English Went Thither*:

"...they turned natural fools upon it for several days: one would blow up a feather in the air; another would dart straws at it with much fury; and another, stark naked, was sitting up in a corner like a monkey, grinning and making mows [grimaces] at them."

"In this frantic condition they were confined, lest they should, in their folly, destroy themselves—though it was observed that all their actions were full of innocence and good nature. Indeed, they were not very cleanly; for they would have wallowed in their own excrements, if they had not been prevented. A thousand such simple tricks they played, and after eleven days returned themselves again, not remembering anything that had passed."

The few who recovered put the blame on an odd, white-flowered plant they'd thrown in the pot. Thus, Jamestown weed was named. Over time, the name was tangled up into 'jimsonweed'. And so the name has remained, to this day.

It's no wonder that making soup (or salad?) out of the plant caused such a fuss. All parts of the plant contain dangerous levels of the alkaloids atropine, hyoscyamine, and scopolamine. The seeds and roots are particularly potent. If you get the dose just right, it's a hallucinogen. If you get it even a tiny bit wrong, you are very sick and probably dead. Jimsonweed poisoning occurs in most domesticated production animals: Cattle, goats, horses, sheep, swine, poultry and people. In animals, it usually happens when the seeds end up in food, or when the foraging isn't good in a field and a plant is growing there that a cow takes a nibble on. Humans aren't as smart as animals; they keep trying to experiment with the plant to get high. And they die from it.

Early Signs of Poisoning:

- rapid pulse
- restlessness
- depression
- rapid breathing
- nervousness
- dilated pupils
- muscular twitching
- frequent urination
- diarrhea
- anorexia
- weight loss

Fatal Cases of Poisoning:

- weak pulse
- irregular breathing
- lower body temperature
- coma
- retained urine
- Convulsions

Let me repeat: it is *not* a plant to trifle with.

It arrived long enough ago that nobody is completely sure how it came to North America. Some say that it was traded among the Native American tribes, working its way up the continent as a powerful and perilous medicine. Others say it was brought by the Spaniards and traded as a narcotic with the French and English settlements. But all we can do today is speculate about this strange, spiny bloom.

If You Love It: It does have its charms. The creamy white blooms are fragrant at night, attracting moths and other nocturnal pollinators. Its flowers have earned it many accolades, and have even been immortalized by Georgia O'Keefe. "When I think of the delicate fragrance of the flowers, I almost feel the coolness and sweetness of the evening," the artist was said to have remarked on the subject.

As long as you break off and throw away the seed pods before they open, the plant can be contained and cultivated as a witchy accent piece. Just keep it far, far away from children, pets, and experimental fools. Don't let it seed. And don't get it on your skin.

If You Loathe It: then your main focus is on the seeds. Do not let it seed. Luckily, that is the biggest thing to worry about… aside from the poison, of course.

For The Home

- **Dig:** Wearing protective clothing, dig the plant up and burn it, or if you prefer, send it to a landfill.
- **Dispose**: Don't compost it; you don't want to risk spreading poisons on your garden. It does not regrow once it's yanked, thankfully.
- **Mulch and Replant:** Now for the important bit: preventing next year's seeds from germinating. There are definitely still some left in the soil from other years. My first suggestion: Mulch. Mulch the site at least 2 inches deep, but 3 is better.

The Chemical Option: If you still have issues, you can go chemical and spread a pre-emergent herbicide on the ground. Don't bother with herbicides on the adult plant; the size and the amount of chemical you'd need makes it worse than useless.

Good luck!

Kudzu

Pueraria montana

Common Names: Ko-hemp, kuzu, Japanese arrowroot, Chinese arrowroot, porch vine, telephone vine, mile-a-minute, foot-a-night, cuss-you vine, the vine that ate the South.

Family: Fabaceae
Genus: *Pueraria*

Looks Like: a climbing, deciduous vine with alternating leaves made of three oval-shaped or lobed leaflets. These vines are capable of reaching lengths of over 100 feet in a single season. Its fleshy taproots can reach 7 inches in width and grow to 9 feet in length. These roots can weigh up to 400 pounds! After 3 years, kudzu blooms from July through September. The fragrant, pealike purple and red flowers are produced on plants that are climbing or draped over vegetation or other objects; vines rarely flower when trailing on the ground. Smelling of grape soda, the flowers hang in clusters, suspended from the axils of the leaves. Flowers are followed by flat, hairy seed pods containing 3 to 10 seeds. The seeds mature on the vines in October and November. A thick seed coat protects them, and they can remain viable for up to 10 years when they're buried in the soil.

Comes From: Southeast Asia.

Likes To Live: in open, disturbed areas such as roadsides, right-of-ways, forest edges, and old fields.

States Present: North as far as South Dakota and west as far as New Mexico. It's also been spotted in Washington and Oregon

States Impacted: Any state where it grows. Kudzu's economic impact is enormous. One recent U.S. government publication estimates Kudzu causing over 100 million dollars of damage a year. Another source, which factors in 336 million dollars' worth of lost productivity in forests, estimates total productivity losses to Kudzu as "greater than $500 million per year" (Miller).

Legend
- List only
- Law only
- Both List and Law
- Included on other source
- Not Listed

Succeeds By: growing a foot a day until it reaches maturity at three years of age. After that, it spreads underground by runners; vines growing along the ground can root every foot or so at the nodes, and form new root crowns that become independent plants. And each one of those plants, as it matures, sets down its own runners, flowers, and drops its own seeds. Up to 30 vines may grow from a single plant. Between its growth, the amazing storage of nutrients in its tuberous central roots, and its mile-a-minute runners, it's impossible to control.

Causes Trouble By: using everything and anything as a scaffold for its own growth. And I do mean anything. Telephone wires, which can snap with the weight of the vines. Buildings, which it grows over and prises apart at every joint and crack. Trees, which it scrambles up and smothers to death. Railroad tracks, which it grows across. When trains run over the vines, the slick pulp causes the train wheels to slip and sometimes derails the trains. Kudzu overpowers other plants by blanketing leaves, breaking branches, or uprooting entire trees with the weight of the vines. And it keeps on climbing. Abandoned cars? No problem. Old junkyards? Great. And, according to rural lore, it'll climb over people who stand still too long… or so they say.

Kudzu. The problem child of the horticultural world. The weed that anyone points to when they want to say that government projects are worse than useless. Its story is as twisting as its vines. And it begins with three men, and the biggest show of their day.

In celebration of America's 100th anniversary of independence, the Centennial Exhibition in Fairmount Park opened in May of 1876. Officially named the International Exhibition of Arts, Manufactures, and Products of the Soil and Mine, it took up 285 acres, took in 10 million guests, and ran until November. 37 nations participated in the event. The grounds contained 5 major buildings: the Main Exhibition Building, Memorial Hall (Art Gallery), Machinery Hall, Agricultural Hall, and Horticultural Hall.

Inside the Horticultural Hall was, among other wonders, a Japanese pavilion with a house and garden. In this pavilion was a variety of beautiful new plants from Japan being brought into Western cultivation. Twined up the porch posts of the little home was a plant with lovely purple blooms. Viewers were enchanted. A newspaper gushed that it "grows a mile a minute in the face of all difficulties."

Three men were responsible for the plants on show. Samuel B. Parsons Jr. was an innovative landscape architect and nurseryman. Around 1839, he founded Parsons and Company Nursery. The Parsons nursery specialized in the acclimatization and introduction of plants from Asia. The 1876 Exposition was, for Parsons, a chance to show just how good his plants could be. There was pride here for two of his major suppliers as well: Thomas Hogg, adventurous florist and nurseryman, and George Rogers Hall, a physician who turned to botany and became one of the foremost plant traders in the business.

Between the three of them, they introduced everything from Japanese Maples to PeeGee hydrangea. They're also responsible for kudzu, sweet autumn clematis, Oriental bittersweet and Japanese knotweed between them.

So, the lovely vine had been introduced. Now both the nurserymen and the Japanese consulate promoted kudzu—"the wonder vine"—as a plant capable of taking root on land that couldn't be cultivated for anything else. This makes sense in the light of modern understanding: kudzu, a member of the legume family, is a nitrogen-fixing plant. It can provide its own nitrogen on poor soil, and improves the soil by adding nitrogen to it.

With its broad, layered leaves and wisteria-like purple flowers, it soon gained popularity as a shade plant and became known as 'porch vine'. It was treated as a pretty, slightly overactive ornamental, the way we treat Virginia creeper today. As an ad in *Good Housekeeping* in 1890 put it, "This is the most remarkable hardy climbing vine of the age, and one that should be planted by every one desiring a dense shade. It flourishes where nothing else will grow, in the best or poorest soil, and owing to its hardy nature, requires little or no care."

In the early 1900s, Florida nursery operators Charles and Lillie Pleas discovered that animals would eat the plant. In 1905, the couple opened Glen Arden Nursery in Chipley and began touting kudzu as the "coming forage of the South," with the help of nature-oriented organizations like the Audubon Society. A mail-order business, they sold kudzu seeds and seedlings through the mail. A historical marker there proudly proclaims "Kudzu Developed Here".

But their modest advocacy was nothing to what came next.

In 1930, during the Great Depression, Georgia experienced the worst drought in state history. The state's soil, already depleted by cotton and tobacco monoculture, was absolutely exhausted. Erosion was rampant. The saying of the time was that the wealth of the South was washing away.

In 1935, the newly formed Soil Conservation Service decided to recommend kudzu for erosion control and soil improvement. They began producing seedlings at nurseries in Virginia, North Carolina, Alabama, and Georgia. Between 1935 and 1942 the Soil Conservation Service grew 100 million kudzu seedlings, shipped them throughout the Southeast and handed them out to farmers. Hundreds of young men were given work planting kudzu, through the Civilian Conservation Corps. Farmers were paid as much as eight dollars an acre as incentive to plant fields of the vines in the 1940s.

The plant had an incredible cheerleader in the person of Channing Cope. A journalist and radio show host in Georgia, he founded the Kudzu Club of America, which eventually had a membership of 20,000. He became known as the Father of Kudzu, constantly advocating—indeed, cheerleading—for the plant. During the 1940s, he traveled across the Southeast starting Kudzu Clubs to honor what he called 'the miracle vine.' "Cotton isn't king in the South anymore," he often crowed, "Kudzu is king!"

Cope lived long enough to see his king dethroned: the government stopped advocating the use of kudzu in 1953. Sources say that Cope didn't end his life well. But the plant he worked so hard to bring to prominence? Oh, it flourished. Without human cultivation—read, interference—the plant was rampant. Without the constant work of farmers cutting it for hay, grazing animals on it, and working it into the soil, it grew undisturbed. And grew. And *grew*. It was declared a weed in 1972, and a noxious invasive in 1980. In 1993 a study by the Congressional Office of Technology

Assessment claimed that kudzu accounted for about 50 million dollars annually in lost farm and timber production. By then, it was thoroughly entrenched, and happily munching away on the South.

If You Love It: ...Well, it's not useless. It does improve the soil. Philip Gant and Yota Batsaki of the *Scientific American* do have thoughts on its uses: '*The Book of Kudzu: A Culinary and Healing Guide*', by William Shurtleff and Akiko Aoyagi, recorded the plant's uses in Japan—for food, cloth and medicine—and proposed their adoption on American soil.

A 19th-century Japanese treatise on kudzu, Seikatsu roku, recently acquired by the DDumbarton Oaks rare book collection in Washington, D.C., reveals just how valuable kudzu was in its East Asian habitat. Written by agricultural innovator Ōkura Nagatsune (1768–c.1860) and illustrated by a student of the famed printmaker and painter Hokusai, the treatise celebrates kudzu as a "useful thing ... in useless places," able to flourish in depleted soils and steep mountain sides. The book brings together detailed methods and instructions for collecting and processing the plant for food and textile manufacture. The author appeals to regional administrators to scale up kudzu production as a safeguard against famine and to promote innovative regional applications as a spur to economic development.

Ōkura is well aware of kudzu's role in traditional Chinese medicine, where the root extract has long been used as a treatment for alcohol intoxication as well as hypertension, coughs and colds. Modern science is beginning to produce clinical evidence of puerarin's effect against excessive alcohol consumption and alcohol damage. Other studies suggest that the plant and its extracts may inhibit HIV-1 entry into cell lines, remove toxins from the ground and build biohybrid circuits that contribute to solar power in the developing world.

Ōkura might also take heart at the imaginative uses of the plant in East Asia today. In Korea (where since at least the 17th century agricultural manuals included kudzu in their instructions for famine relief and country living) the plant is used in over 100 different products, from face masks to prepackaged cold noodles. And in Japan, it is an essential element of regional specialty items ranging from "Kakegawa handbags to Kumamoto medicinal extracts to crispy kanemochi." (Gant, Batsaki)

If You Loathe It: Oh boy. Have you got reasons. The plant is a disaster; about the only thing that really kills it dead is cold. It already infests 9 million acres of land. And with global warming really cranking up the temperature, if anything its range is set to increase. But some control measures have been figured out. Kudzu bug, arriving in 2009, has begun to help control the plant. Unfortunately, the bug also attacks soybeans. And it's a true stinkbug. So… it isn't as helpful as all that. Here's the basic steps to controlling kudzu:

For the Home

- **Find a vine lead:** Grab a couple of the vines and walk towards where you feel resistance. This will tell you where the kudzu crown is located.
- **Cut the vines:** You'll find intertwined clusters of vines. Get into them, and start cutting. I recommend starting with shears, or a hand pruner for small infestation. Just make sure your blades are sharp. The plants can be composted, or tossed into a mulcher or a landfill. They quickly wither without their roots, which is a blessing.
- **Remove soil-attached nodes and small crowns:** As you pull along the vines and cut them, you'll notice knot-like areas on the vines, every 6 to 16 inches. Every one of these can become a new root crown. Get rid of them; either dig them up, or cut them at the ground.
- **Get at larger crowns:** You'll need to find all the well-established crowns, and clear the ground around them. Dig down a couple feet to find where the crown meets the taproot; at the place where the plant material goes from pale to dark, you can stop digging.
- **Sever the crown from the tap root:** Once you have enough room in the hole for tools, cut the crown from the roots using loppers, shears if easy enough, or—my personal favorite—a reciprocating saw. Blessedly, at this point the kudzu keels over.
- **Mulch and Replant:** This is essential! Get new, desirable plants in the ground in this site and care well for them. Mulch of at least 2 inches deep is the best way to prevent any kudzu seeds in the soil from sprouting.

For Larger Areas: Regular monthly mowing can kill kudzu eventually, but you have to be persistent with this. Grazing is also an especially good way to handle the infestation. Goats are basically mowers with hooves, and I'd say they'd be the best way to go.

The Chemical Option: While glyphosate can kill the plant and the Missouri Department of Conservation Invasive Species Coordinator recommends applying clopyralid herbicide to kudzu during the summer, I say don't bother unless your infestation is *very* small. The amount of herbicide you'd need to kill very much kudzu will kill everything else too, and poison the soil for some time.

Mullein

Verbascum thapsus

Common Names: common mullein, flannel plant, candlewick plant, high taper, hag's taper, velvet dock, velvet plant, lung-wort, Our Lady's candle, Our Lady's flannel, Our Lady's blanket, Jacob's staff, Aaron's rod, Adam's flannel, Adam's staff, Saint Peter's staff, sow's ear, Indian tobacco, velvet dock, velvet plant, rabbit's ears, old man's blanket, shepherd's staff, Quaker's rouge, white man's footsteps.

Family: Scrophulariaceae
Genus: *Verbascum*

Looks Like: a softly elegant biennial. First year plants develop as a basal rosette of felt-like, gray-green leaves. Basal leaves are 4 to 12 inches long and 1 to 5 inches wide, covered with woolly hair. The leaves grow alternately from the stem, decreasing in size towards the top. The plant bolts in the second year, creating a stalk 5 to 10 feet in height. Flowering occurs in June to August, when 5-petaled, yellow flowers develop at the top 3 to 5 feet of the stalk. The flowers are also self-fertile, with self-fertilization occurring when the flower closes at the end of the day if cross-pollination hasn't happened. This means that a lone plant won't be alone for long. The fruit is an ovoid capsule that splits, releasing 100,000 to 180,000 seeds. Plants die after flowering, leaving a tall, thin brown monument of a stem.

Comes From: Northern Africa.

Likes To Live: in well-drained soils. It prefers dry, sandy sites but can grow in chalk and limestone. It can be found in neglected meadows, forest openings, pastures, fence rows, roadsides, old fields, construction zones and industrial areas.

States Present: All 50. It's well into Canada too.

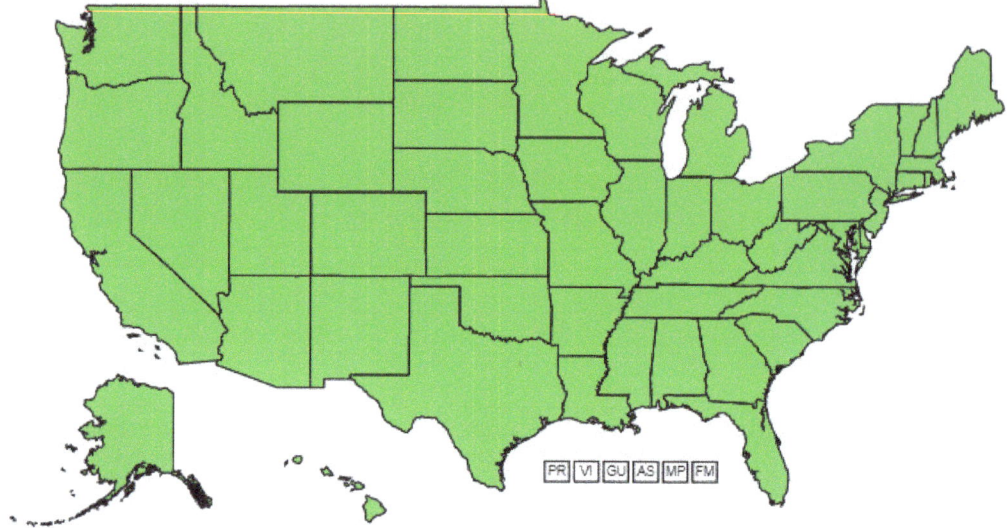

States Impacted: Eastern states, California.

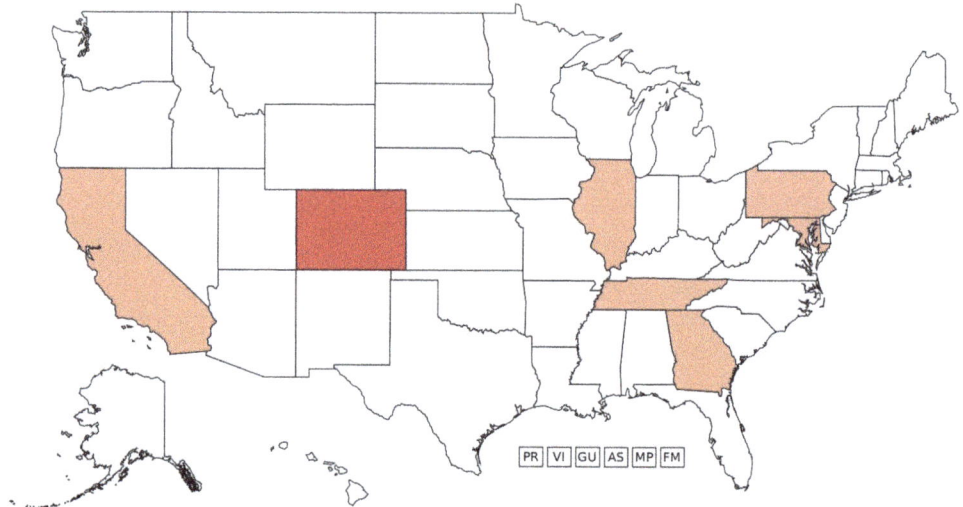

Legend
- List only
- Law only
- Both List and Law
- Included on other source
- Not Listed

126

Succeeds By: seeds. It seeds prolifically and spreads the seeds prodigally. Every stalk of blooms produces 200 to 300 capsules, with 500 to 800 seeds per capsule. That brings seed production to somewhere around 100,000 to 240,000 seeds per plant. Seeds in the ground can be viable for 80 years (!)

Causes Trouble By: outcompeting native plants. Once established, it grows quickly to form a dense ground cover. At high densities, it eats up the space and nutrients that native plants need, especially in areas prone to fire; these plants move faster than the natives to establish themselves, beating them to the punch and disrupting normal ecological progression following fires or other disturbances. It also serves as a host for insects that are economic pests, such as the mullein leaf bug, a pest of apples and pears in the Eastern United States and Canada.

Mullein has been a part of the Western world from the beginning. To the Greeks, it was *fluma* or *flego*, both words meaning 'the flame'. Dipped in fat, suet, oil or tallow, it was used as ready-made torches, tinder, and wicks for candles. The Romans, as usual, copied off Greek homework and called the plant *caldelaria;* 'the candle'.

The word 'mullein' comes from the Middle English moleyne and the Old French moleine, originally from the Latin *mollis*, meaning 'soft'. One of its popular names, "lungwort," derives from its most common use: from ancient Rome to modern Ireland, a tea made from its leaves has been used as a cure for lung diseases in both humans and livestock.

Hunters and fishermen also had a use for it. Aristotle was the first to record the use of mullein as a fish poison. The seeds were ground up and tossed into a pond or slow-moving stream. Chemicals in the seeds leached into the water, were absorbed through the gills of the fish, and caused them difficulty in breathing. The fish eventually floated to the surface where they could be easily scooped up for supper.

With so many uses, it's no surprise that the plant was introduced. The Puritans brought mullein seeds to America for their medicinal herb gardens. By the late 1630s, mullein had escaped to neighboring fields and roadsides. As settlers moved west and planted new gardens, patches of mullein marked every abandoned homestead. Records show that it was first described in Michigan in 1839. By 1859, mullein's persistence outside of cultivation was beginning to earn it a bad reputation. "There is no surer evidence of a slovenly, negligent farmer, than to see his fields over-run with Mulleins" writes William Darlington in his *American Weeds and Useful Plants*, published that year. It was recorded on the Pacific coast in 1876. Today, it often seems to live up to its old name in its blooming season, when it seems to set entire hillsides alight in yellow flares.

If You Love It: Mullein does have a lot of wonderful uses. When it's grown as an herbal remedy, every part of the mullein plant is usable at different times during its life cycle. This is a plant I know well from my own childhood; a warm oil infused with mullein leaves and flowers was dripped into my ears to cure childhood earaches, and it was very effective. The thick, soft leaves were used as bandages on small cuts and scrapes in the fields around my childhood home, and they worked very well if you used a little milkweed sap to stick them on. The leaves, boiled into a tea and drunk before bed, were the best thing to make sure you shed the nasty cough you'd had for a bit by morning. And this makes sense, now that we know more about its biochemistry. Chock-full of mucilage, the plant also contains a lot of phenolic compounds, vitamin C, vitamin E, carotenoids, and terpenes, making it a very good anti-inflammatory and antibacterial, especially for the sensitive mucus membranes. Lab studies have shown mullein leaf has antibacterial properties in both Gram-positive and Gram-negative bacteria, including Klebsiella pneumoniae, Escherichia coli and staphylococcus aureus.

This makes it a good choice as something to take for an annoying but harmless cough. Do drain the tea through a coffee filter though; the tiny hairs on the leaves can be irritating. Mullein flowers can provide a soothing skin wash, too. Use mullein as a wash or balm for minor skin wounds and acne by infusing flowers in olive oil and adding beeswax to make a balm.

Speaking of beeswax, the bees absolutely adore mullein. If you choose to, plant mullein in an area where bees can enjoy the flowers, and birds can enjoy the seeds. Just make sure you cut them before they seed all over, or wrap the seed heads in a bag so that you can gather the seeds and decide where you want them to go. If you choose to forage wild mullein, only harvest from straight, vigorous stalks that have not been treated with herbicide.

If You Loathe It: at least it's not one of the difficult plants to handle.

For the Home

- **Bag it:** Before you do anything else, put a bag over the flower spikes so you don't spill the innumerable seeds everywhere.
- **Pull it:** If you do not want to be overrun with mullein, remove the rosettes quickly in their first year.
- **Dig it:** The adults can be dug out of the ground prior to seeding. The root is a bit like a carrot, and not too difficult to get out.
- **Give it a haircut:** Cut down the flower stalk before it seeds. The plant will die, and it won't be any more trouble.
- **Watch out for it:** The seeds are still in the soil, and will try to sprout for decades. I recommend mulching the ground and/or encouraging a strong and healthy growth of groundcover. After that, just keep an eye out.

For Larger Spaces: Mullein can't survive being mowed or tilled, so these are both perfectly good options. Grazing by goats is also another good choice.

The Chemical Option: On mullein, herbicides are really overkill. But in the case that you really feel the need, go for 2,4 D; it's effective and low-danger.

Pigweed

Amaranthus spp.

Common Names: Pigweed is a name that covers several species in the genus Amaranthus, some native to America and some not. These include:

- redroot pigweed (*A. retroflexus*)
- smooth pigweed (*A. hybridus*)
- Powell amaranth (*A. powelii*)
- Palmer amaranth (*A. palmeri*)
- spiny amaranth (*A. spinosus*)
- tumble pigweed (*A. albus*)
- prostrate pigweed (*A. blitoides*)

They're known by lots of names, including red-root amaranth, redroot pigweed, red-rooted pigweed, common amaranth, pigweed amaranth, and common tumbleweed.

Family: Amaranthaceae
Genus: *Amaranthus*

Looks Like: tall, erect-to-bushy plants anywhere from 2 to 5 feet high, with simple, oval-to-diamond-shaped, alternate leaves. The leaves of some Palmer amaranth plants have a whitish V-shaped mark on them. The flowers are small, green and crowded into coarse, bristly spikes at the top of the plant. Smaller spikes are located in the leaf axils below. Each plant bears thousands of flowers. As warm-season annuals, they emerge, grow, flower, set seed, and die within the frost-free growing season. Female flowers form small, round, shiny dark reddish-brown to black seeds, roughly a quarter of an inch across. Ninety thousand seeds weigh one ounce.

Comes From: mostly, South America, with a couple subspecies from Eurasia.

Likes To Live: anywhere it can find some sun and some water. It isn't picky. Disturbed areas like fields, yards, and the edges of woods are just fine. Construction sites and drainage ditches work too. It puts up with all soil types, but it really thrives in the soil of a well-amended garden.

States Present: Everywhere but Georgia and South Carolina.

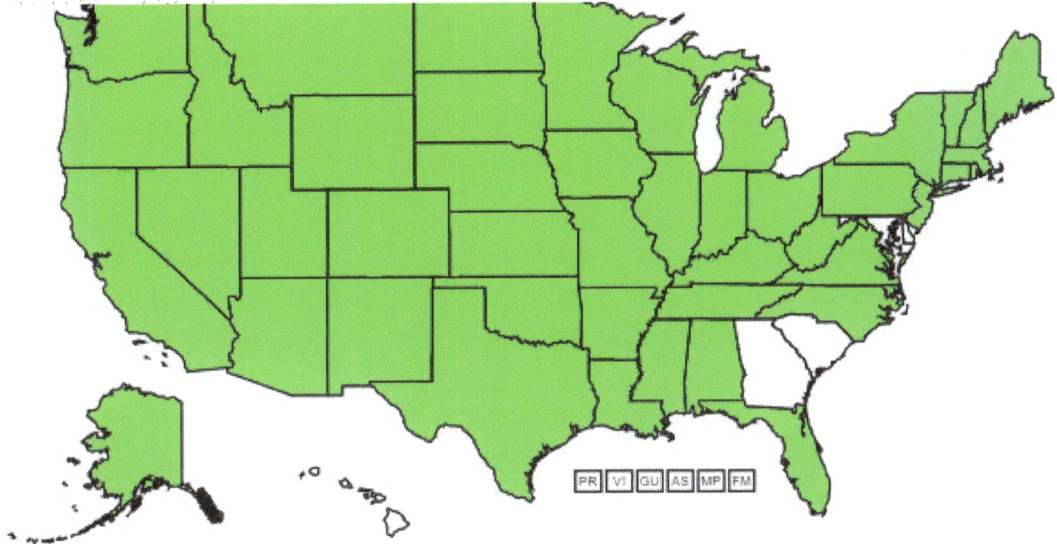

States Impacted: Every one where it grows.

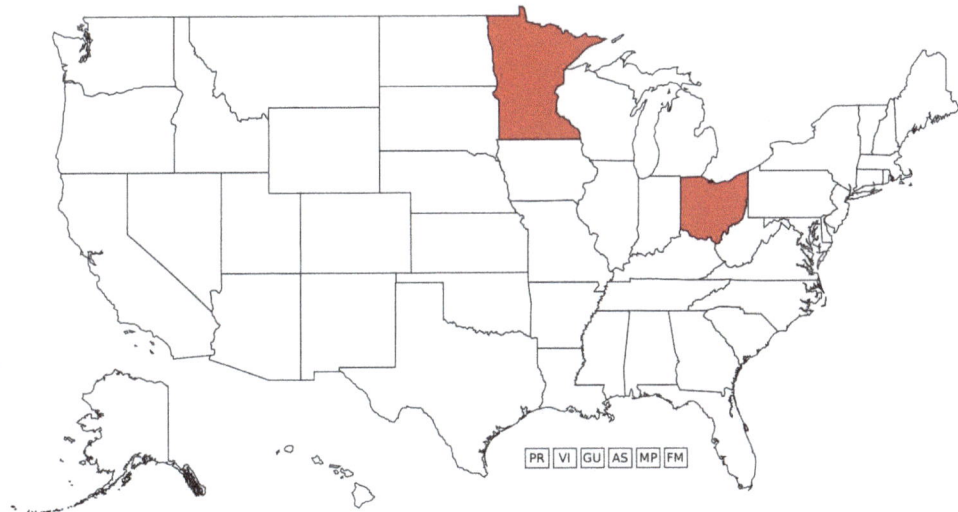

Legend
- List only
- Law only
- Both List and Law
- Included on other source
- Not Listed

133

Succeeds By: spreading seeds. Everywhere. All the time. A single large plant can bear 100,000 to 600,000 seeds. Oh, and they all hybridize. All the pigweeds are cross-fertile, and most of them are self-fertile as well.

Causes Trouble By: outcompeting economically valuable plants. Pigweeds thrive in hot weather, tolerate drought, respond to high levels of available nutrients, and are adapted to avoid being shaded out through speedy stem elongation. They compete aggressively against warm season crops, and seed faster than you'd believe. The plants grow about 2 to 3 inches a day and reach heights of 6 to 8 feet, beating crops to the light and inhibiting crop growth. They're hungry weeds too, sucking the soil dry of nutrients.

Economic thresholds for pigweed in soybeans and corn vary from 5 to 15 plants per 30 feet of row, depending on the crop and its price. Yield losses have been up to 91% in corn and 79% in soybean. The loss can be almost 97% in cotton. Pigweed is also a dangerous forage contaminant, the plants building up nitrates in their stems and branches in concentrations high enough to poison livestock. It's harmful and even deadly when fed to cattle and pigs, destroying their kidneys. Infestations in silage corn have been reported to kill cattle.

Pigweeds serve as alternate hosts for crop pests as well: the green peach aphid, tarnished plant bug, European corn borer, flea beetle, cucumber mosaic virus, and strains of Fusarium and Rhizoctonia that attack sugar beet.

And the really bad news? Many of the pigweeds have become glyphosate resistant due to the way it's been sprayed. The species is considered a threat specifically to genetically modified cotton and soybean crops in the southern United States, because the plant has developed resistance to glyphosate since at least 2006. Glyphosate-resistant pigweed not only dominates the cotton fields, it impacts all sorts of other crops and productions. Uh-oh.

In the days before Cortez put his armored and blood-stained feet on South American land, the four main crops in the valley of Mexico were centli (maize), huauhtli (amaranth), etl (beans) and chia. Archeological records have given us a sense of how important these plants were; In

Central America, domesticated amaranth seeds were recovered from Coxcatlan cave in the Tehuacan Valley of Mexico, in contexts dated to 4000 BCE. These four plants were the life of Central and South America. And they were honored for it. Statues of these plants were made for the deities. Idols of dough were created each year, made from amaranth and toasted corn seeds mixed with honey or maguey sap and called zoale. The zoale were made in the month of May in honor of the war god, Huitzilopochtli. During the festival, these zoale were broken up, shared and eaten all through the community. The Aztecs used amaranth during at least six other seasonal festivals honoring their gods. The Aztecs also made tortillas and tamales out of amaranth flour along and used the greens in daily life as well.

Given the way the Spaniards in this time period thought, it should be no surprise that huauhtli was banned by the invading Spanish Empire in 1519. Amaranth fields were burned and cultivators were punished. Lucky for us, they were unable to completely destroy the grain. In a few remote areas, amaranth survived. The sacred staple food became an outlaw and a renegade.

As humans hunted and neglected it, huauhtli learned new tricks. It grew smaller and less colorful. It lurked in the corners of the fields it had once been lovingly tended in. It got into the crop and its seed was mixed in with others. As colonists traded and spread, the seeds spread about as well. Soon, they were finding new ground further up the continent, their tiny seeds spread by birds, animals, and human movement.

Once revered and pampered, the feral huauhtli thrived, sneaking out from under the heels on the throats of its people and taking off for new pastures, literally. The *American Indian Health and Diet Project* writes,

> "As agricultural land was expanded in the Americas by European Americans, pigweeds are thought to have become more common and developed their reputation as an unwanted pest. At the time of initial Euro-American colonization of North America, when only a small portion of exposed soils were present, it is no wonder that the inconspicuous pigweeds were not documented as a common food plant of the Native Americans."

Soon this once-honored plant had earned a new and much less complimentary name: pigweed, a plant that was fit only for feeding to pigs. Thomas G. Fessendon is credited with the first use of the word pigweed in the English language, but it probably applied to *Chenopodium*. In his *Original Poems*, revised edition published in 1806, he wrote:

> "The hyacinth and daffodil,
> With now and then a big weed,
> Of purslain and of pig weed."

More recently, pigweed is up against a new enemy: Monsanto. The company thought they'd hit a home run when they started putting out Roundup Ready plants in the 90s. But nature always bats last. Since 2006, all forms of pigweed have been coming up as glyphosate resistant. In a *Weed Science* journal report of Palmer amaranth's impact on Arkansas cotton fields, researchers determined what would happen if 20,000 seeds—2 percent of the seed potential of a single plant—were released on a square mile test area. The headline was: "One weed leads to complete crop failure."

To me, there's some poetry in what pigweed does these days. Watching feral pigweed, canny and clever plant of the oppressed peoples, go after the cosseted, patented and genetically modified cotton of smug oppressors, I will admit it: I smile grimly.

If You Love It: Pigweed is a wonderful wild food. For the casual forager, the leaves are a lot less work than the seed. The greens are similar to spinach, beet greens, and chard—they're all in the same family, Amaranthaceae—but pigweed has more than twice as much vitamin C as kale and 4 times more than spinach. It's also high in vitamin A and calcium, as well as a host of other vitamins and minerals. Just make sure you get it from a source where it won't have been sprayed, clean it well, and don't eat it raw; the oxalic acid in the leaves can cause kidney stones in large amounts. For greens, pick the leaves from young plants. Smaller leaves are more tender and more nutrient-rich than their bigger relations.

When sources say 'amaranth' they usually refer to amaranth seed, and most of the published nutrition information refers to the seed, or amaranth flour. The tiny seeds contain eight to nine grams of protein in a 1 cup serving, offering a nutritionally complete plant food that has all the essential amino acids needed by the human body, without gluten.

The seeds are ready to harvest when they start falling off of the plant, usually in late summer. Cut off the entire flower head and put in a paper or mesh bag, and then let them dry for a week or two in a well-ventilated spot out of the light. Attics and the tops of fridges are perfect. Shake the seeds loose and pour them into a jar. Well worth the work!

If You Loathe It: then watch your timing, work hard, and weed, weed, weed!

For the Home

- **Grab the fire wand:** A handheld fire wand is a great approach for killing the young plants. Use caution, and play the fire over the plant until it crisps.
- **Mulch, mulch, mulch:** Don't let that gigantic seed bank see the light!
- **Monitor and pull:** The plants come out of the ground easily.
- **Do NOT let it seed!** It will take about 6 years to exhaust the seeds in the soil. But if you miss a plant, you're back at square one.

For Larger Spaces

- **Mow:** It won't kill the plants, but it will help control their growth.
- **Till and cover crop:** Tilling the plants under and immediately planting winter wheat or rye can defeat the plants. These crops can hold their own against the pigweed.
- **Goats:** The oxalic acid doesn't hurt them, and they can take down pigweed faster than any mower.

The Chemical Option: ...You can try it. If you want to watch a plant laugh at you, that is.

Plantain

Plantago major

Common Names: Waybread, waybroad, mother of herbs, cuckoo bread, cuckoo's bread, snakeweed, rabbit's ears, rabbit's tail, Englishman's foot, white man's footsteps.

Family: Plantaginaceae
Genus: *Plantago*

Looks Like: a demure cool-season perennial with broad, oval leaves, up to 4 inches long and 3 inches wide. Smooth or slightly hairy, the leaves have a waxy surface and veins that are parallel to the margins, which are un-toothed and sometimes wavy. The flowers are produced on unbranched stalks that arise from the rosette. They rise between 5 and 15 inches high, clustered with small flowers that have whitish petals and bracts surrounding the flowers. Overall, the flower spikes look like tiny green cattails. The fruit is a 2-celled oval capsule, 2 inches long, that opens by a lid around the middle. The seeds are absolutely tiny, about 1/50 to 1/25 of an inch long. The roots are fibrous and shallow.

Comes From: Eurasia. Can we get more specific? Well, not anymore.

Likes To Live: anywhere it finds a bare spot. It pops up in lawns, along footpaths, in vineyards, orchards, gardens, construction sites, parking lots, playgrounds, landscaped areas, roadsides, and other disturbed locations. It can be found on compacted and soggy sites where other plants will not thrive. Essentially, it's one of nature's ways of trying to heal wounded and degraded land.

States Present: Every one of them.

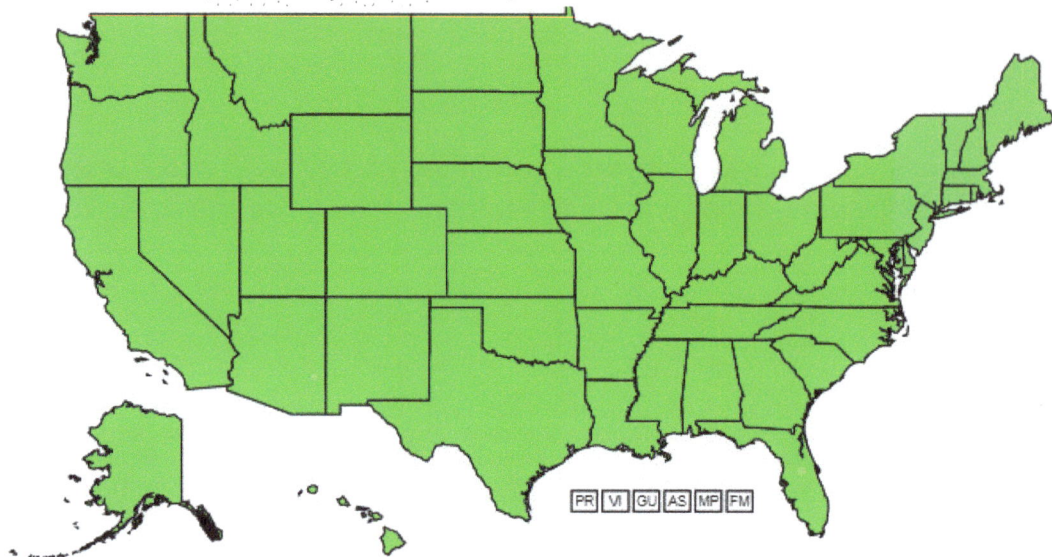

PR | VI | GU | AS | MP | FM

States Impacted: None in particular; it's not harmful.

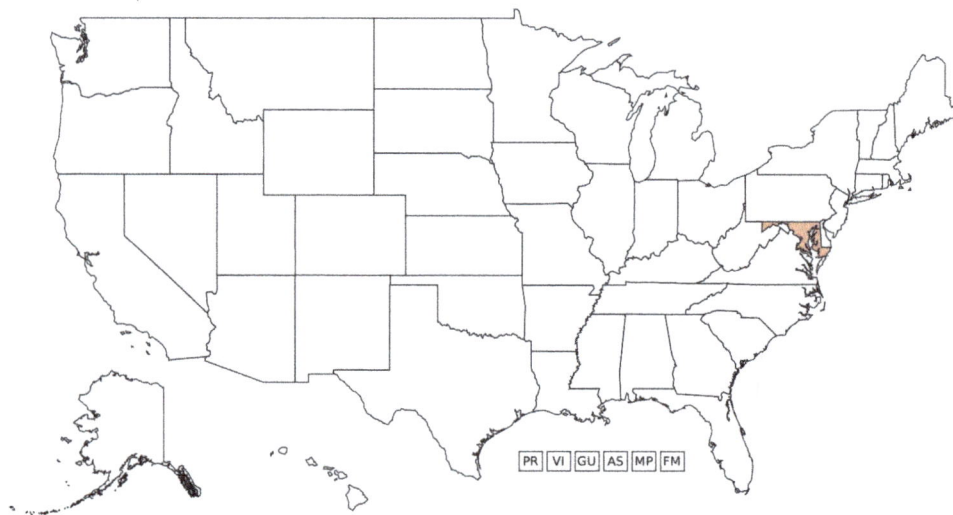

Legend
- List only
- Law only
- Both List and Law
- Included on other source
- Not Listed

PR | VI | GU | AS | MP | FM

140

Succeeds By: spreading prodigious numbers of tiny seeds in midsummer. These seeds are famously tough and can remain viable in the soil for up to twenty years. And the crown is amazingly resilient. Mow it, step on it, weed-whack it, it's all the same to plantain. It simply regenerates and goes about its business.

Causes Trouble By: getting HOAs and property managers in a snit. As a general rule, plantain is no problem at all. Athletic fields aim to keep broadleaf plantain presence low to prevent bad footing. In lawns, where weed cover is more of an aesthetic issue, it's really up to the homeowner. Since I find the plantain more valuable than the grass, I'd say it's not really causing any trouble at all.

> Ond þú, Wegbráde, wyrta módor,
> éastan openo, innan mihtigu;
> ofer ðé crætu curran, ofer ðe cwene reodan,
> ofer ðé brýde bryodedon, ofer þé fearras fnærdon.
> Eallum þú þon wiðstóde and wiðstunedest;
> swa ðú wiðstonde áttre and onflyge
> and þæm laðan þe geond lond fereð.
>
> And thou, Waybread, mother of worts
> open to the east, mighty within;
> over thee carts creaked, over thee queens rode,
> over thee brides cried out, over thee bulls snorted.
> All of them thou withstood and dashed against;
> so may thou withstand venom and that which flies
> and the loathsome that yond the land fareth.

These words, from the thirteenth-century Anglo-Saxon book *The Lacnunga*, show you how much this plant was loved in past days. It is a plant of the roadside and the wayside, the traveller and the track. Its tiny, slightly sticky seeds have traveled on the bare feet, shoes and wheels of people, dogs, horses and carts down the ages. It has traveled everywhere with humans, been mixed in our food, our grain, and our stories. It was mentioned by Shakespeare in a joke.

> "Romeo and Juliet" (i. 2)
> "Benvolio. Take thou some new infection to thy eye,
> And the rank poison of the old will die.
> Romeo. Your plantain leaf is excellent for that.
> Benvolio. For what, I pray thee?
> Romeo. For your broken skin."

It's always been with us, whether we've been paying attention to it or not. And so it is with us today.

If You Love It: Then you've got good sense! A use for it that I grew up with is the oldest and the simplest: when dealing with a scraped knee or a bug bite, grab a handful of plantain leaves, smush (or chew) them, and place the result on the injury. This does a great deal to soothe small wounds, stop the itching from insect bites and stinging nettle, and start the healing. Holding it on with a sticky milkweed leaf is always a good idea.

For more sophisticated uses, consider the chemical constituents of the plant. Studies have found that Plantago major is effective as a wound healer, as well as an antiulcerative, antidiabetic, antidiarrhoeal, anti-inflammatory, antinociceptive, antibacterial, and antiviral agent. It contains a number of effective chemicals, including flavonoids, alkaloids, terpenoids, phenolic acid derivatives, iridoid glycosides, fatty acids, polysaccharides and vitamins.

If you're a foraging type, definitely find some plantain! The plant contains beta-carotene, crude fibre, dietary fibre, fat, plenty of protein and carbohydrates. It also contains vitamins A, B1, B2, B3, C, and K along with a bagful of trace minerals: calcium, chromium, iron, magnesium, manganese, phosphorus, potassium, selenium, and zinc. Good nutrients all the way around! Make sure you're foraging in safe areas that have not been sprayed. After careful rinsing, young leaves can be eaten raw or cooked. They have an earthy, slightly grassy flavor with undertones of pepper. The roots are a bit milder. Because older leaves tend to be tough, they're better suited to cooked dishes.

As a part of a household medicine cabinet, plantain is worth its weight in gold. Medically, its most active chemical constituents are aucubin, allantoin and mucilage. This makes it particularly good for soothing and easing coughs. A classic mix is succus: half plantain juice and half honey. Take it by the teaspoon for nasty coughs and sore throats. A tea of the leaves is good as a skin wash for acne, rosacea, and other inflammations. Oh, and it's good for your digestion too. Ever heard of psyllium, mucilaginous bulking laxative extraordinaire? It's plantain seeds. Mix them in water and drink for those sorts of days.

If You Loathe It: Then it's not particularly difficult to remove.

For the Home

- **Give the lawn the advantage:** Maintain proper soil pH: 6.0 to 6.8. Test your soil every 3 to 5 years. Fertilize according to soil test results and at the proper time during the growing season. Irrigate as needed, deeply and not too often, to avoid creating shallow, weak roots. Mow at proper height, removing no more than 1/3 of the blade at a time. Choose the right turfgrass seed for your conditions, buy quality seed, and overseed the thin spots in fall or early spring. Most importantly, aerate the lawn to prevent compaction problems, and overseed all the bare spots.
- **Dig it:** As long as you pop the plant's crowns out of the ground, you win the fight. A weed-popper or soil knife is more than sufficient. The key is to get it out of the ground before it seeds. Once it seeds, it's a whole new ballgame.
- **Burn it:** For control of this plant, a fire-wand is your friend. Being a cool-season plant, plantain is very sensitive to heat and has no defenses against fire. Alternately, dump boiling water on it; that'll kill it dead.
- **Mulch it:** A good layer of mulch will prevent the plant from getting going.
- **Use it!** With all its uses, you can control it by turning it into part of your diet. Just yank it for the kitchen whenever it gets in the way.

For Larger Spaces:

- **Outcompeting:** Seed or plant a strong native grass or groundcover to beat plantain at its own game.
- **Acceptance:** Since it really isn't doing much damage, let it be.

The Chemical Option:

Plantain is susceptible to all weed-killers. Use a selective herbicide; nonselective herbicides are really overkill. Generally, I'd say herbicides are overkill all together with plantain, but if it's a specialized area like a playing field, then a selective herbicide will do just fine.

Puncturevine

Tribulus terrestris

Common Names: Caltrop, goathead, goat's head weed, cat's head, bullhead, Texas sandbur, Mexican sandbur, devil's thorn, devilhead, tackweed. The other things it's commonly called are not polite.

Family: Zygophyllaceae
Genus: *Tribulus*

Looks Like: a warm-season annual herb growing along the ground, coming up from a woody taproot that can be 9 feet deep. The leaves are opposite, oblong and short-stalked, 1 to 3 inches long, hairy, and pinnately compound (having leaflets). Each leaflet is 1/4 inch long. The stems are numerous, hairy, and up to 6 feet long, forming a dense mat. The small, yellow flowers have 5 petals, small and pretty while they last. Each flower lasts only a few days, but the plant blooms intermittently from midsummer to hard frost.

The fruit that follows each flower's bloom is called a 'nutlet'. The fruit separates at maturity into 5 bony seed casings, each of which is pyramidal and tipped at the corners with rigid spines like needles. They're strong enough to puncture bicycle tires and go through the sole of your shoe. Within each casing, seeds are stacked on top of each other, separated by a hard membrane. A typical puncturevine plant will produce up to 5,000 seeds during the growing season. Which is why they're called a great many very rude things.

Comes From: Africa and the Mediterannian.

Likes To Live: Hot, dry, barren spots where few other plants can survive.

States Present: All the states of the West, along with a handful of benighted Midwestern states.

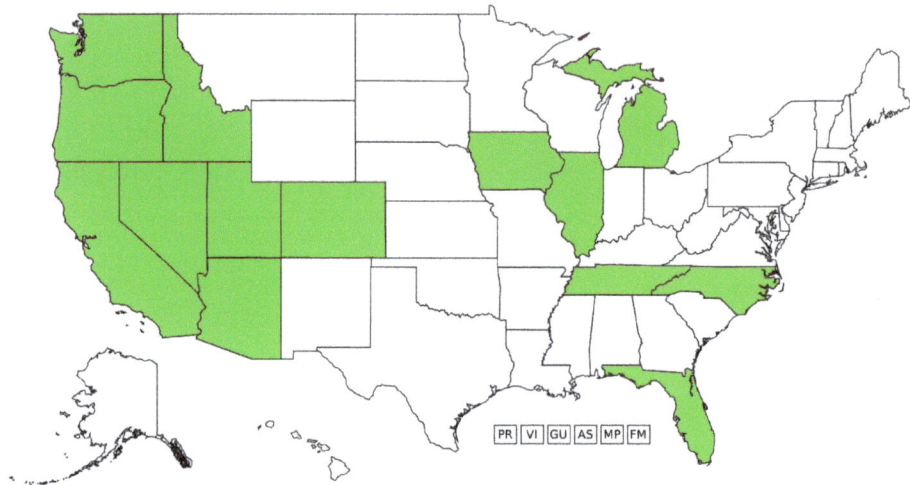

States Impacted: Puncturevine is regulated as a state noxious weed in Arizona, California, Colorado, Idaho, Iowa, Michigan, Nevada, North Carolina, Oregon, South Dakota, and Washington. It's regulated as a Class B Noxious Weed in North Carolina. Yeah, it's a problem.

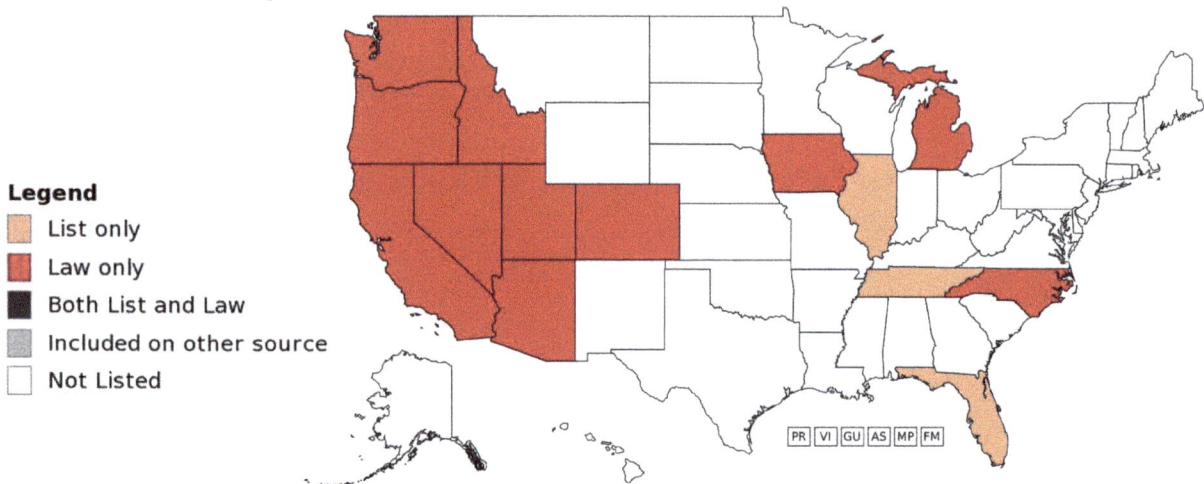

Legend
- List only
- Law only
- Both List and Law
- Included on other source
- Not Listed

Succeeds By: being a little… ahem. To put this politely, being a nuisance. Excuse me, this plant has stabbed my feet many, many times. Its sharply thorned seeds can dig into and stick on any surface, living or not. They're produced at an incredible rate, they stab into and stick to anything that they touch, and they can float when what little water there is arrives.

Causes Trouble By: interfering. With everything. Everywhere it grows. It causes trouble with harvesting; the nutlets are a nuisance to pickers in orchards, vineyards and gardens. They can also contaminate produce, particularly dried fruit and alfalfa. In its common habitat in the compacted soil of roads and paths, puncturevine is notorious for puncturing bicycle tires, shoes, boots, and the feet of pets and people. If the thorns break off, they become the source of infections, and can cause sepsis and gangrene in humans and animals alike.

Puncturevine is even more dangerous in pastures. Livestock can suffer injury to their mouths, eyes, digestive tracts and skin from the nutlets. Painful punctures of the feet happen on a daily basis, sometimes causing infection and lameness, especially in horses. In badly infested pastures, wounds to the mouths of animals can cause the animals to avoid eating and lose condition. If that's not bad enough, it's toxic; mainly sheep, but goats and cattle too can get very sick. At least 3 forms of poisoning are involved: nitrate poisoning can happen when sheep eat the very young plants. The nitrate is converted to nitrite in the rumen, leading to death. Photosensitization results after animals eat the plant when it has fungus, making sheep listless and sometimes causing death; and the staggers (ataxia) arise if very much of the mature plant is eaten. About 20,000 sheep died of eating the plant in Australia during the drought of 1981-83.

In case that wasn't bad enough, puncturevine is reported as an alternative host of root-knot nematode, bean leaf roll virus—causing stunt in chickpea—and tomato spotted wilt virus. Oh, and silverleaf whitefly and sugarcane weevil. Sounds lovely, doesn't it?

For such a painfully dramatic plant, the story of its introduction is fairly prosaic. Might we even say pedestrian? I know, terrible pun. I couldn't resist.

It was introduced to North America by early European settlers unintentionally; the plant's nasty, thorny seeds made their way across the ocean from Spain, embedded in imported sheep wool. It arrived in California around 1900, and spread rapidly along railroads and highways. It moved incredibly fast, thanks to the hitchhiking skills of its seeds. Puncturevine was found in the Pacific Northwest, growing in the gravel along the railroad tracks in Whitman County, Washington, in 1924. And now we're stuck with it.

Nobody likes goatheads. The Latin name *tribulus* originally meant 'caltrop', a type of spiky weapon dropped to damage attackers' feet. Even in Roman times, the word already meant this plant as well.

Although its extract is used as a dietary supplement with the belief that it increases testosterone levels mainly for body-builders, controlled studies have proven that it doesn't work and might be dangerous.

If You Love It: ...might I suggest you reconsider?

If You Loathe It: Let's talk about control measures.

For the Home

- **Get It Early:** Puncturevine spreads by seed, so killing the plants before they seed is absolutely essential. If you've been through an infested area make sure to clean your shoes, clothing and tires to prevent spreading seeds to other areas. Use any of these, or a combination of these methods to control puncturevine. For small infestations, use a hoe to cut the plant off at its taproot. Luckily, it can't resprout from the root. For large specimens, cut the plant off at the soil surface with a hoe or pruners, and bag it. Sweep the nutlets and loose seeds up with a broom. Get every one that you can. Throw the whole mess away, or burn the lot. Please don't compost it. It won't end well for anyone.
- **Bury It Deep:** Covering the area with a deep layer of amended topsoil, followed by mulch, can provide better ground to plant in and prevent existing plants from germinating both.

A seed that can't see light is a seed that doesn't sprout.
- **Replant!:** After puncturevine is under control, plant areas with native plants that can shade the ground and outcompete the little monsters.

For Larger Spaces:
Really, the only sensible things are chemical or bio-control in larger spaces.

The Biological Options:

- **Puncturevine Weevils:** The perfect predator for the notorious puncturevine. Puncturevine weevils—actually two species with the Latin monikers of *Microlarinus lypriformis* and *M. lareynii*—work well in low-traffic areas that have enough puncturevine to keep the weevils alive. *Microlarinus lypriformis* larvae burrow into the spiny seed to lay their young. *M. lareynii* larvae eat the stems. Imported in 1961 from Europe specifically to start handling puncturevine, it's an efficient wee beastie. When the larvae hatch, they literally devour the plant and its seeds from the inside out. There's a delicious schadenfreude in releasing the beetles to get after puncturevine.
- **Outcompeting:** Overseeding large areas with a crop that beats it to the punch is a great control method. Buffalo grass, native wildflowers, barley, winter rye, and winter wheat are all good choices.

The Chemical Option: A solid pre-emergent, applied for three years, will clear the existing seed bank out. Applying pre-emergent herbicides like trifluralin, chlorsulfuron, or dichlobenil in late winter will control germinating seeds of puncturevine. After germination, imazapyr, dicamba and 2,4-D will also reduce infestations.

Contact your local county extension agent for recommended use rates, locations, and timing. Once it's controlled, the issue is making sure it stays out.

Purple Dead Nettle

Lamium purpureum

Common Names: Archangel, Gill-go-by-the-ground, henbit (shares this name and many traits with *Lamium amplexicaule*), red dead nettle, henbit deadnettle, cat's foot, turn-hoof, field-balm.

Family: Lamiaceae
Genus: *Lamium*

Looks Like: a cool-season annual in the mint family, it has the classic square stems of the mints. These stems reach between 6 and 15 inches in height. The plant has a fibrous root system. The leaves are somewhat rounded, coarsely toothed, and opposite.

The flowers are about 1-1/2 inches long, purple, and held in hairy bracts, blooming early in the spring. When blooming has finished, the seeds are formed in nutlike fruits of a varying brown hue.

Comes From: Europe. We can't be more accurate than that, these days.

Likes To Live: wherever the soil has been disturbed and there's plenty of water. Fields, gardens and areas along buildings, spots under trees, yards, parks, roadsides, fallow fields, winter grain crops, pastures, and turf grass.

States Present: It's easier to list the states that don't have it: Minnesota, the Dakotas, Wyoming, Nevada and Arizona. It does like water, after all.

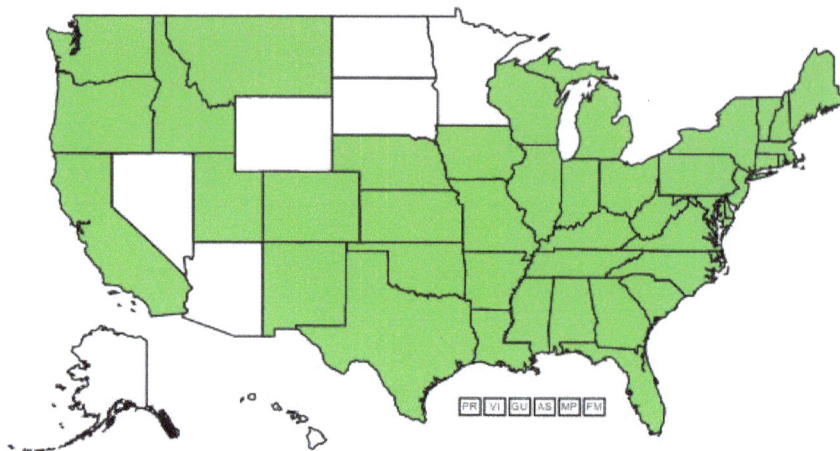

States Impacted: Anywhere growing soybeans.

Legend
- List only
- Law only
- Both List and Law
- Included on other source
- Not Listed

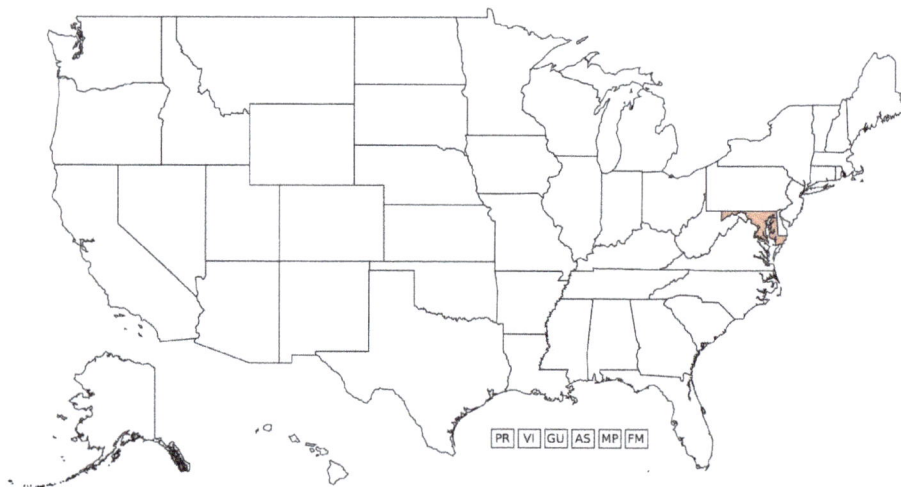

Succeeds By: being quietly persistent, and seeding prolifically. Purple deadnettle has been found to produce more than 27,000 seeds per plant, its fragments have the ability to root wherever they touch, and its runners can spread it for miles. A winter annual, it germinates in the cool days of autumn, grows until hard frost, overwinters and then gets a headstart on everything else in early spring. It's already flowering in March and April and seeding in May, before most plants have even woken up properly.

Causes Trouble By: meandering into any little nook or cranny with enough shade and water for it, by the use of its long runners. It's prevalent in winter cereal crops, other winter annual crops, orchards, gardens, landscapes, nurseries, waste areas, and turf; more commonly warm-season than cool-season turf, which it can outcompete in early spring. It's not really a problem, but it can degrade the aesthetics of a property by outcompeting grass in shady spots and giving lawns and gardens a scruffy appearance. A winter annual, it gets a headstart on desired plants early in the spring, and can vie with them for nutrients early in their lives, which makes it hard for them to get established and thrive. In disturbed natural sites, it can replace the natural vegetation and threaten biodiversity.

It causes enough trouble in spring agricultural fields to get some states to declare this pretty little weed as a noxious pest. It's been found to be an overwintering host for soybean cyst nematode and corn earworm. The soybean cyst nematode is a serious threat to soybean crops, so it is a serious danger for farmers and those working with soybeans. In fact, Purdue University researchers found that in the greenhouse, SCN reproduction on purple deadnettle and henbit was so efficient that it often equaled or exceeded SCN-susceptible soybean. Purdue research also showed SCN juveniles were present inside purple deadnettle and henbit roots in both fall and spring. However, juveniles were more abundant in spring weed infestations.

"It makes the head merry, drives away melancholy, quickens the spirits, is good against quartan agues, stancheth bleeding at mouth and nose, if it be stamped and applied to the nape of the neck; the herb also bruised, and with some salt and vinegar, and hog-grease, laid upon a hard tumour or swelling, or that vulgarly called the king's evil, do help to dissolve or discuss them; and being in like manner applied, doth much allay the pains,

155

and give ease to the gout, sciatica, and other pains of the joints and sinews. It is also very effectual to heal green wounds, and old ulcers; also to stay their fretting, gnawing and spreading. It draweth forth splinters, and such like things gotten into the flesh, and is very good against bruises and burnings. But the yellow archangel is most commended for old, filthy, corrupt sores and ulcers, yea, although they grow to be hollow, and to dissolve tumors."

Culpeper, the writer of the quote above, gives you a sense of how much the plant was appreciated in merry old England. This was also where it picked up the name 'dead nettle'; it gained the name because it has a passing resemblance to stinging nettles, without the nasty punch that plant can pack. Hence, it came to be known as the 'dead' nettle.

How it arrived in America is a mystery today, but the general assumption is that purple deadnettle's seed mixed in with the seeds of crops it loves to live beside, and arrived to be broadcast along with barley and winter wheat. Today, it's become one of the signs of spring, its gentle persistence promising warmer days ahead.

If You Love It: then you and I are on the same page. This is a nuanced little plant; yes, there are places where it cannot and must not be. But kept out of harm's way, it has many charms. It's an invaluable food for pollinators early in the season, when they are the most vulnerable. Bees love purple dead nettle; it's an important species for them, almost as important as dandelions. If you live in a mild climate, it may even flower through the winter. Small birds also love its seeds.

It's pretty good for you, too. Purple dead nettle contains polyphenols, compounds with antioxidant properties that are associated with a broad range of health benefits. It's also a source of quercetin, a flavonoid with immunostimulating and anti-inflammatory properties that is used to ease allergies and other chronic inflammatory issues like rheumatism. It's full of Vitamins A and C, as well as being a good source of iron. The flowering tops can be finely chopped and dropped in salads or cooked dishes like soups, sauces, egg dishes, and stir-fries. Dead nettle's somewhat fluffy texture lends itself especially well to fritters. It has a mild, slightly grassy, somewhat floral flavor, and the purple tops are even a little sweet. Cooked, it can be used in any recipe where you would use spinach.

Traditionally, it's particularly useful for drawing splinters and infections out of a wound as a hot poultice, and its polyphenols back this use up.

If you like it but don't want to encourage a weedy species, there are a number of cultivated dead nettle species that are lovely.

If You Loathe It: then you have some work ahead of you.

For the Home

- **Give the lawn the upper hand:** Growing a thick, healthy lawn is the first line of defense against these mint cousins, since the grass will easily outcompete the weeds for nutrients and growing space when it's healthy. If you've got a spot in the yard where the nettle beats the grass, find out the reason why. Sometimes the shade a tree casts or a low spot that holds water makes it hard for the grass. This is when you need a special grass blend.
- **Give it a tug:** You can yank these shallow-rooted flowers easily to make way for your spring planting. But be warned, you'll have to do this for at least a few years.
- **Give it time:** If you find it fairly harmless or if it's in a fairly empty part of the garden, you can let the little plants alone and enjoy a few weeks of their gentle purple colors. When the weather warms up, the plants will go dormant and wait for the cool days of autumn.

For Larger Spaces:

- **Till:** The plant is not deeply rooted, and can't survive tilling.
- **Cover Crop:** If the ground isn't open, the dead nettle doesn't do well.
- **Go chemical:** A pre-emergent is very effective.

The Chemical Option: Post-emergent herbicides that contain metsulfuron or trifloxysulfuron can be used against purple deadnettle coming up in Bermuda grass, but pre-emergent herbicides are much safer for other grasses. Be sure to apply pre-emergent herbicides in the late fall or early winter, before the purple deadnettle starts to germinate.

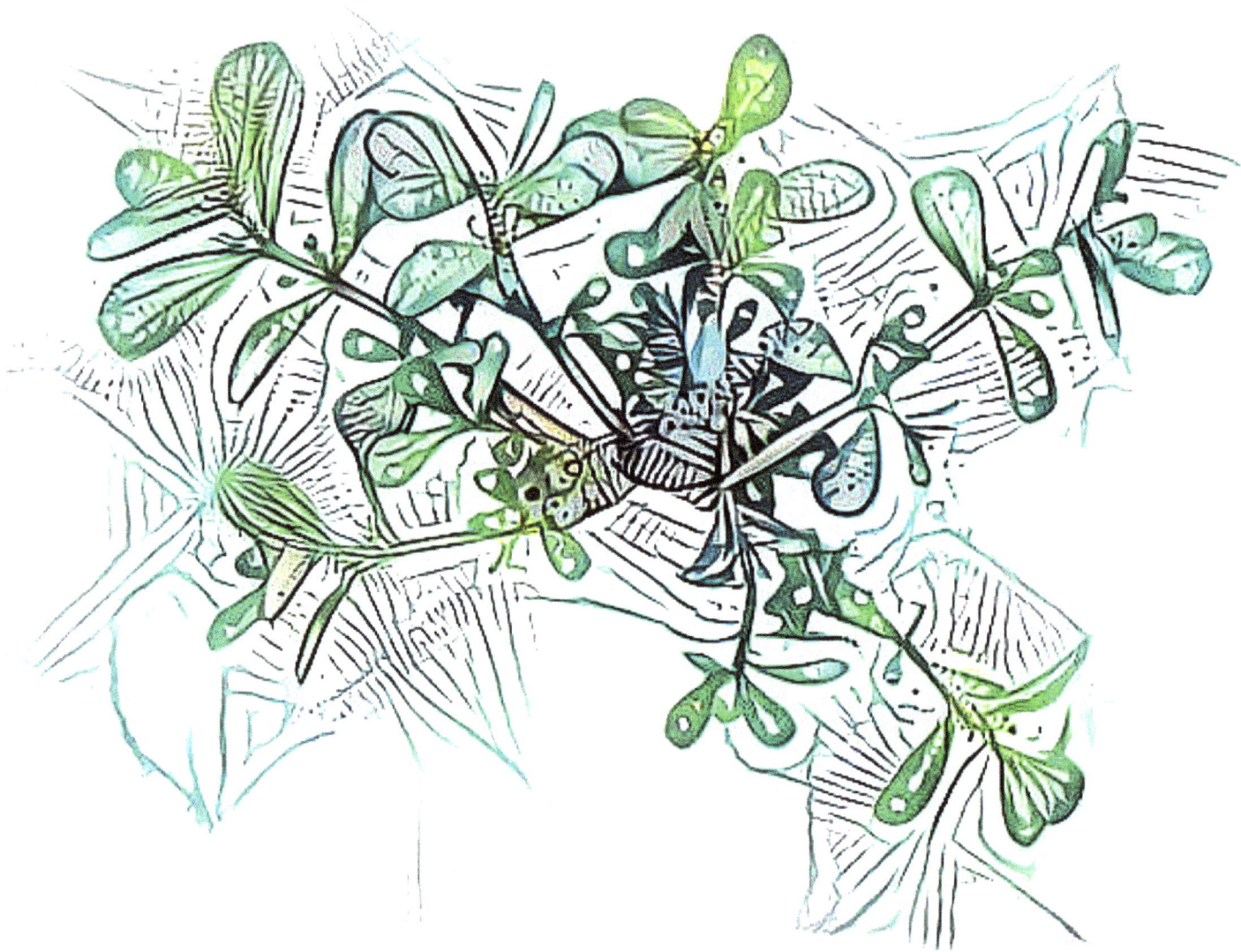

Purslane

Portulaca oleracea L.

Common Names: little hogweed, courpier, pourcellaine, pourpie potager, pourpier, akulikuli-kula, common purslane, garden purslane, duckweed, purslane, pursley, pusley, snackvine, snackweed, snapweed.

Family: Portulacaceae
Genus: *Portulaca*

Looks Like: a summer annual that forms a spreading mat, 6 inches tall and up to 2 feet across, branching frequently at the base. The stems are round, thick, and succulent, smooth and reddish. The leaves can be alternate or opposite, up to an inch long and fleshy, with a rounded egg or spade-like appearance. The yellow flowers have 5 regular parts, about 1/4 inch wide. Depending upon rainfall, the flowers can appear at any time in the growing season. The tiny seeds are formed in a pod, which opens when the seeds are mature. It has a shallow taproot with fibrous secondary roots.

Comes From: a place that is somewhat debatable. Archeology has proved that some branch of the plant's family tree is native to Canada and was clearly eaten as part of the Salts Cave inhabitant's diet in the pre-Columbian days of Kentucky, but sources agree that another branch of the family was brought over from Southern Europe, to be used as a pot-herb by settlers in the 17th century. So, let's call it a tossup for now.

Likes To Live: everywhere from rich garden soil to roadsides, from trampled ground to parking lots and sidewalk cracks. Just give it sun and don't overwater it.

Every one of them, and it's all over Canada and Middle America too.

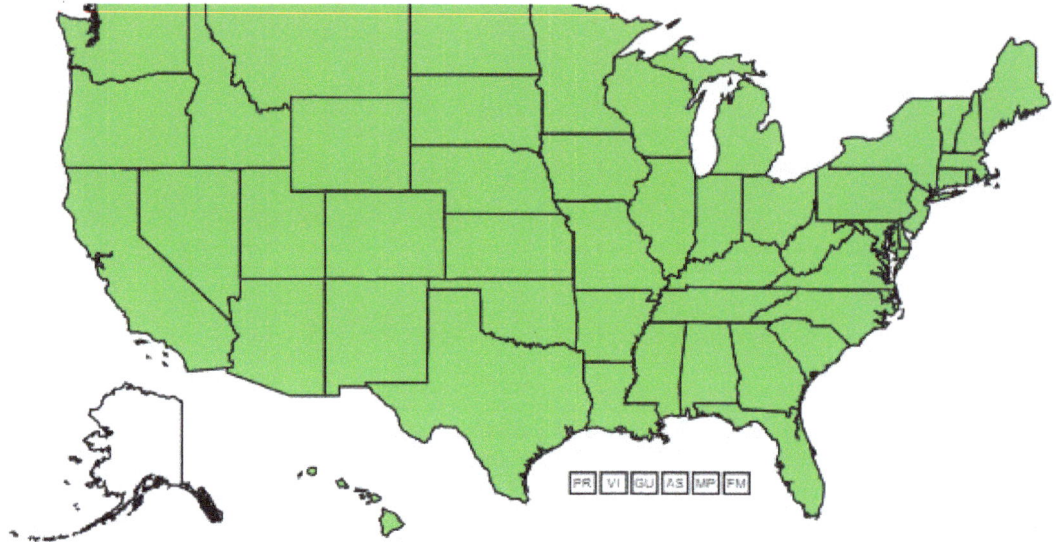

States Impacted: Purslane is noted as one of the 10 most noxious weeds worldwide. Any state with a lot of agricultural fields and heat has a problem with purslane.

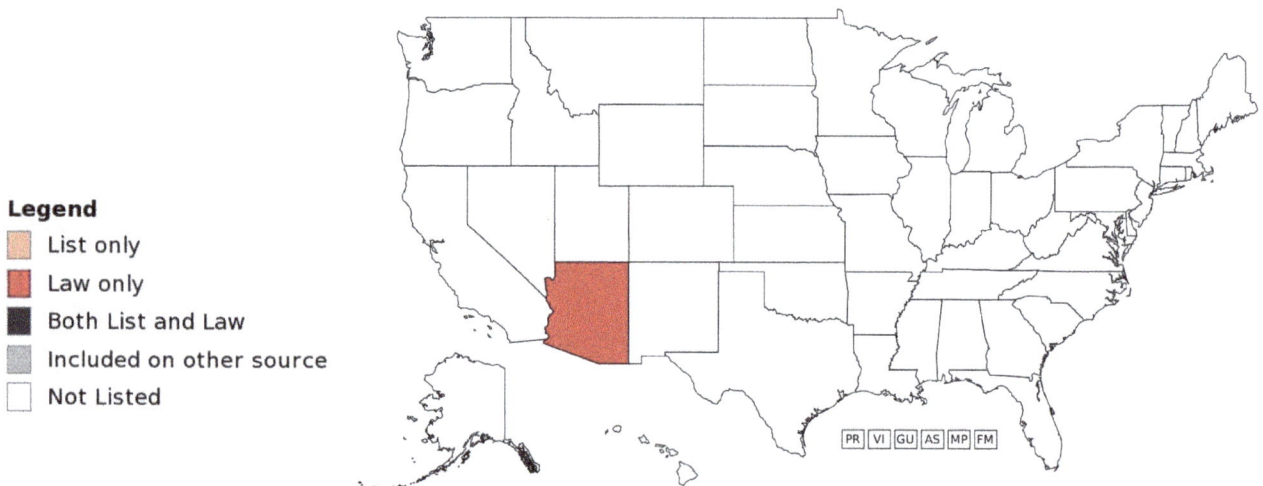

Legend

- List only
- Law only
- Both List and Law
- Included on other source
- Not Listed

162

Succeeds By: seed production, survival and drought tolerance. Purslane is one of very few plants able to utilize both CAM and C4 photosynthesis pathways, methods of photosynthesizing that were once thought to be incompatible with each other. Purslane will switch from C4 to CAM pathways during droughts, which makes it incredibly tough. The seed survival is amazing: the seeds can remain viable up to 19 years in dry storage, and as long as 40 years when buried in soil. And there's an awful lot of those seeds. A single plant may produce 240,000 seeds in a season. Because the plants are self-fertile, a few scattered plants in the first year can become a solid carpet of purslane the next.

Causes Trouble By: getting in where it isn't wanted and making the place look scruffy. In agricultural areas, it competes for resources with field crops, particularly species that are germinating or growing in midsummer. Affected crops include: asparagus, red beets, celery, crucifers, cotton, maize, onions, potatoes, rice, soybeans, sugarcane, tomatoes and wheat. Since the purslane is better at retaining water, it survives when other plants struggle.

Purslane is one of those plants that has been with us since the beginning. Archaeobotanical work has retrieved purslane seeds from a Stone Age site in Ontario; the speculation is that this plant was brought for food by the ancestors of the Native Americans, in their crossing of the Bering Strait. There is evidence that purslane has been eaten for at least 2,000 years; it was cultivated in ancient Egypt and was enjoyed by the ancient Romans and Greeks. Ancient Greek doctors found purslane helpful in treating inflammation in the urinary system, mouth, digestive tract, and many dietary ailments. It was known to the Arabs in the medieval period, and may have been cultivated in Europe as early as the 13th century. The 17th century monk Agapius Landus of Crete suggested a fresh green salad made with purslane, basil, rocket, cress, and garlic to those suffering the common cold or 'the ague'. Purslane was reported in Massachusetts as early as 1672, which makes sense; it was treated as a healthy and useful addition to the diet in those days. And no wonder!

Modern researchers found that purslane is one of the very few plants that contain alpha-linolenic acid, a type of omega-3 fatty acid normally found in fish. It contains an explosion of vitamin C, some vitamin B and carotenoids, as well as minerals such as magnesium, calcium, potassium and iron. Given the average diet in 1672, purslane could be a literal lifesaver. Tough and incredibly drought-tolerant, purslane patiently, calmly keeps on growing, whether we're paying attention to it or not.

If You Love It: Eat it! It is a rich source of potassium (494 mg/100 g) along with magnesium (68 mg/100 g) and calcium (65 mg/100 g) as well as that all-important omega-3 fatty acid. It is very good source of alpha-linolenic acid (ALA) and gamma-linolenic acid (LNA, 18 : 3 w3) (4 mg/g fresh weight). The antioxidant content and nutritional value of purslane are wonderful. Snap the branches off and snack on them in areas where you know it hasn't been sprayed. Add it to salads for a bit of color. In Turkish cuisine purslane is used just like spinach, though it's a bit brighter and more lemony.

One warning: don't confuse purslane with two-spotted spurge, a much more dangerous but similar plant from a different family. Two-spotted spurge has flat leaves and, when broken, exudes a milky sap. Don't eat that, and wash your hands immediately; the sap isn't good for you.

If You Loathe It: Then it's very easy to deal with.

For the Home

- **Give the lawn the advantage:** In healthy lawns, purslane doesn't stand a chance. It only wins the game in parched, starving turf. So! Irrigate as needed, deeply and not too often; you want to avoid creating shallow roots vulnerable to the heat. Mow at proper height, removing no more than 1/3 of the blade at a time. Or go with my favorite choice: dig up the lawn, mulch the site and plant a native garden; less water needed, more ability to beat the weeds at their own game.
- **Dig it:** A weed-popper or soil knife is more than enough. Purslane comes out of the ground very easily. The trick is to get it out of the ground before it seeds. Once it seeds, you have much bigger problems.
- **Burn it:** A fire-wand is your friend any time purslane shows up in cracks in the hardscape.

- **Mulch it:** A good layer of mulch will prevent the plant's seeds from germinating.
- **Use it!** With all its uses, you can control it by turning it into part of your diet. Pull, chop, eat!

For Larger Spaces:

- **Outcompeting:** Seed or plant a strong native grass or heat-tolerant groundcover to beat the purslane at its own game.
- **Tilling:** If you can pull it up for a few successive years, you can get rid of it.

The Chemical Option: Weed killers are usually overkill with purslane. But when it's needed, as in a large turf-grass planting, a granular pre-emergent is effective. Put it down in early spring, around the last week of February or first week of March depending on your local weather. In farm crops, herbicides can work when the plants are young (May, roughly). Once they mature, there's no point; the waxy leaves wick the poison. An issue of Weed Science makes a statement on the subject:

> "Common purslane can be controlled preemergence with metribuzin applied at 11/3–22/3 ounces per acre. Postemergence application of 2,4-D amine at 1 quart per acre or a tank-mix of Callisto (mesotrione) at 3 fluid ounces and atrazine at 1 pint per acre will provide control of common purslane. Cultivation can also control common purslane in row middles; however, stem fragments can root at the nodes if not completely desiccated following cultivation."

In home settings, I'd say just yank it.

Siberian Elm

Latin: *Ulmus pumila*

Common Names: Siberian elm, Chinese elm, littleleaf elm, dwarf elm, Asiatic elm, trash tree.

Family: Ulmaceae
Genus: *Ulmus*

Looks Like: a fast-growing tree, reaching 50 to 70 feet in height. Its leaves are alternate, oblong in shape, 1 to 3 inches long, and usually have serrate (saw-toothed) margins. The branches form a very regular structure, looking like a set of Vs arranged along the central stem. The flowers are greenish and clustered, with short pedicels, and appear with or before the leaves from March through April. The bark is a light gray-brown with irregular furrows, often streaked with stains caused by bacterial wetwood. The fruit, called a samara, ripens from April to May. It looks like a dry, compressed nutlet surrounded by a papery wing.

Comes From: the Gobi Desert of Asia.

Likes To Live: any bloody place. I've seen it grow up right through chain link fences, or coming up through the concrete sidewalk. This didn't end well for the sidewalk or the fence. It's a bit terrifying in its tenacity. It tolerates a wide variety of growing conditions, including extreme temperatures, nutrient-poor soils, city pollution and low moisture. It also grows scattered in moist soils along streams. Siberian elm is resistant to the Dutch elm disease that destroyed our beautiful native elm. It can be found along roadsides, in pastures and grasslands, stream banks and prairies.

States Present: Only the Deep South and Maine have escaped the invasion… so far.

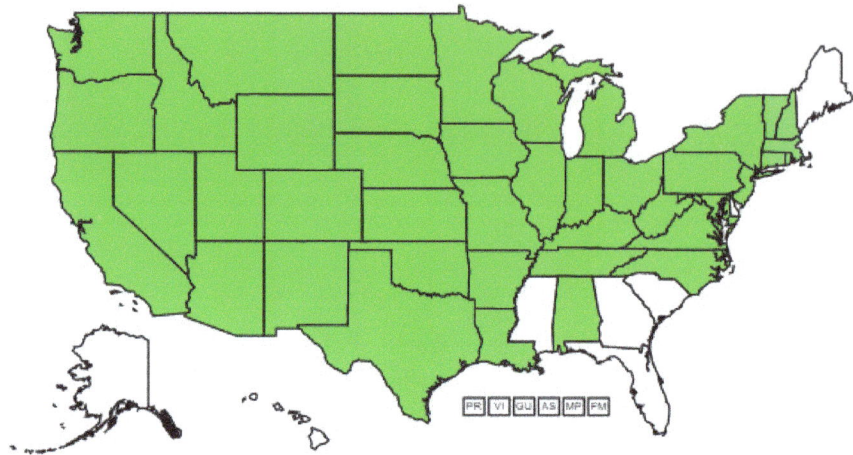

States Impacted: The West is really battling the plant.

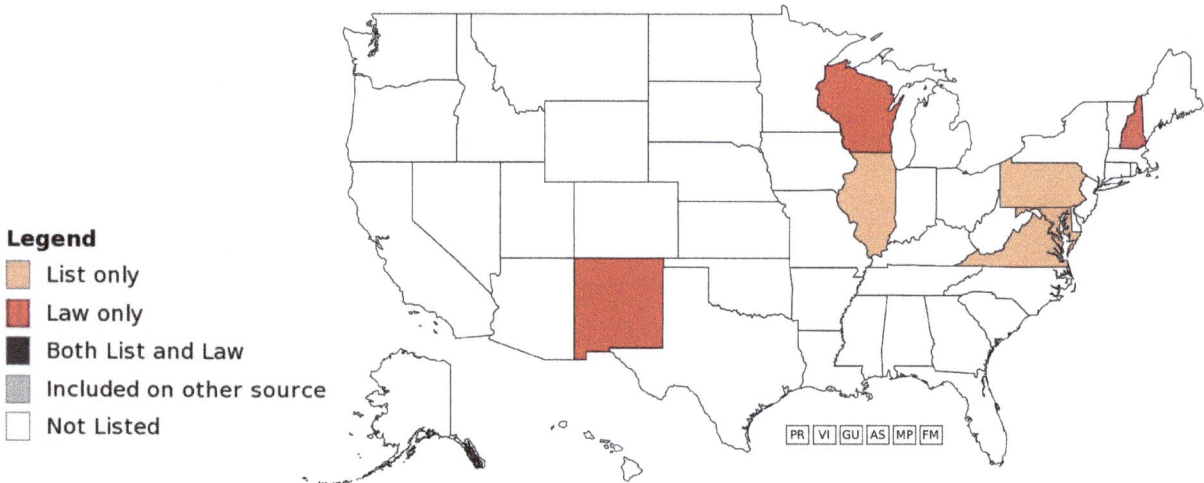

Legend
- List only
- Law only
- Both List and Law
- Included on other source
- Not Listed

Succeeds By: seeding fast, growing fast, and holding its own against drought and heat as well as wind and cold. Siberian elm is an aggressive species that can invade and outcompete the native plants. The tree is able to establish and dominate the prairies in just a few years.

Causes Trouble By: being a terrifyingly determined beast that lives fast, digs deep, and dies hard. Dryland prairies and urban areas everywhere are vulnerable to Siberian elm invasion. Thickets of seedlings form around seed-producing trees, bare ground, animal and insect mounds, construction sites, houses and other disturbed areas. Wind carries the seeds to new areas, where new colonies can form.

This tough interloper survives under conditions that most other species struggle with, allowing it to take advantage of open ground and resources that native plants desperately need. Fast growing seedlings of Siberian elm quickly outcompete the native vegetation, shading them out and starving them of light. And then there's what it does to property.

'Siberian Elm, Not Recommended', reads the Morton Arboretum entry for this plant. 'Not recommended.' No kidding! This is another one of those plants that makes modern gardeners glare at their forebears. Siberian elm was somebody's bright idea. Three specimens were supplied by the famous and well-regarded Späth nursery of Berlin to the Royal Botanic Garden Edinburgh in 1902 as *U. pumila*. Kew Gardens obtained specimens of *U. pumila* from the Arnold Arboretum in 1908, and it became fairly popular in this wet, gray little island; it seemed to be well-behaved, healthy, and tolerant of difficult sites. Difficult *English* sites, that is.

The tree was formally introduced to the United States in 1905 by Professor J. G. Jack of the Arnold Arboretum, when he sent specimens to the Boston gardens in 1905. That same year Frank N. Meyer, an Agricultural Explorer for the Department of Agriculture, and Charles S. Sargent of the Arnold each collected seeds of Siberian elm in Peking. In 1910 E.H. Wilson, another member of the Arnold, collected seeds from this species at the Temple of Peking.

Around 1914, Meyer—he of the infamous Bradford pear—brought a large shipment of the seeds back from somewhere near Peking. Some of the first trees from these seeds were received by the University of California at Berkeley in March, 1917. Because of Siberian elm's fast growth and tolerance of almost any site, nurserymen raved about it, promoted it as a wonder tree, and were responsible for its tremendous success as a street tree. But since the tree had been available in international plant catalogs and had been bought and sold internationally as an ornamental since the 1860s, it was already well known by then. These nurserymen just made the problem worse.

And then came the Dust Bowl. In the 1930s, the plains dried up and cracked, kicking up blinding, mile-wide and mile-high clouds of dirt. Animals died and crops refused to grow.

The tree was cultivated at the USDA Experimental Station at Mandan, North Dakota, where it flourished. After its success at the experimental station, it was selected by the USDA for planting in shelter belts across the prairies in the midst of the Dust Bowl disasters. It was recommended for use and extensively planted during the 1930s and 1940s, to ease the devastation of the Dust Bowl.

According to the periodical *The Santa Fe New Mexican*, a New Mexico governor handed out the seeds of Siberian elm trees to anyone who would plant them. Ben Wright, an arborist and chairman of the Taos Tree Board, was interviewed by the *New Mexican*, and said that Siberian elms are uniquely adapted to the Southwestern landscape, needing little space or water to grow and are able to withstand strong winds. "In the West, it was introduced to solve problems," he added. "As are many troublesome species."

The naturalization of Siberian elm is similar to the earlier naturalization of tamarix (*Tamarix pentandra Pall.*) and Russian olive (*Elaeagnus angustifolia L.*), both of which also became noxious weeds in time.

In the 1950s it began to be sold nationwide on the back of the Sunday paper magazine section as a hedge plant to replace privet. This must be one of the worst horticultural ideas of all time, yet hundreds of thousands of trees were sold and it spread to all corners of the U.S. Its shockingly fast growth and tolerance for drought and cold made it a success... at first. And then its weaknesses began to show, along with its dangerous strengths.

The thousands of wind-dispersed seeds spread the tree. The earliest documented date of establishment in the wild was in Utah, 1935. Dropped into the earth, it germinates and forms dense thickets that choke out the existing plants. Without the native vegetation, native animals don't have their source of food, and they leave.

Siberian elm isn't a long-lived tree either. It has brittle wood which cracks in winter storms. Since it was at one point planted as a shade tree, this means it cracks and smashes open an awful lot of roofs. One of its survival mechanisms is to simply shut off branches and let them die, resulting in mature Siberian elms with crowns that are snarls of dead, dying and living branches like the nests of giant, incompetent birds. These days, with a few exceptions, it's a menace.

If You Love It: It can do a lot of good when properly cared for and watched for trouble, especially down in the deep heat near the Mexican border. It's only going to get hotter down there, and the Siberian elm can survive it. Paul Schmitt, 67, who owns Sunsilk Landscaping in Santa Fe, an area populated with a variety of trees, is proud of his work with Siberian elms. He pointed one out to Kate Moss of *The Santa Fe New Mexican* during their interview. "It's been pruned and cleaned out, and when it is cleaned out they are beautiful trees," he said. Schmitt also has two large Siberian elms in his backyard, protecting his garden from the harsh New Mexico sun. The elms "provide a block from the western sun and cool the yard. They are tough, they are resilient. You have to nurture it to get it to be really beautiful, but when you do, they are just stupendous." He said he has refused to remove them from people's yards on landscaping jobs, and instead has told them, "We are going to clean it up...And they come back and thank me," Schmitt added.

There is one silver lining to the invasion: it's effective as a non-endangered replacement for our native Eastern slippery elm, *Ulmus fulva*. Dried and pounded up, the inner bark makes a wonderful soothing tea for sore throats and aching stomachs. So go ahead and strip the weedy invaders for all the bark they have to offer.

If You Loathe It: Then you and I are on the same page. This is one of the most troublesome plants I deal with as a horticulturist. Here's some tips.

For the Home: Options for controlling seedlings include hand pulling, hoeing or tilling. Small trees can be dug up with a shovel. Larger trees can be girdled between late spring and mid-summer, taking care to remove the outer layer of bark. Girdling will deprive the tree of nutrients. Use an axe, hatchet, or hand saw to make a cut all the way around the trunk of the tree, about 3 feet off the ground. The cut should go through the bark and the cambium, a spongy layer just under the bark, but should not get into the heartwood of the tree. On smaller Siberian elm trees, the cut should be about 1/2 inch deep, and on larger trees it should be up to 2 inches deep. Make a similar incision just above or below the first cut. On small trees, the 2 cuts can be 2 inches apart, while on larger trees they should be 6 to 8 inches apart. This done, the tree will slowly die over a year or two. Be aware, the tree will become more brittle as it dies, so if it lives near your house don't do this; call somebody in to get it taken out professionally. It also might be kinder than a lingering death.

Larger trees should be cut down, but cutting through the entire trunk will often trigger re-sprouting below the cut, and they must be removed regularly for several years. Grinding the stumps is an effective way to stop this. If this sounds like a lot, there are professionals out there who can do this work very effectively. Getting rid of a Siberian elm isn't cheap or fun, but it is satisfying to see it gone. Replace it with a native species, and you'll do the wildlife a favor too.

For Larger Spaces: Thickets can be taken out with a weed hog. For large trees, see above. In many areas, the city/state will subsidize the work. Tractor mounted grubbing tools can be used to drive a blade into the soil, cut roots and uproot trees. Large scale clearing can be done with backhoes or excavators to remove top growth and most roots. Follow-up is needed afterwards to stop sprouts from re-establishing. Follow this work by re-establishing desired plants.

The Chemical Option: One way to prevent re-sprouting after you cut down a Siberian elm is to apply stump sprout preventer on the fresh cut; it will be systemically drawn into the tree roots and kill the root system. Please see the best practices section in the back of the book for details and safety measures. Other than that, professionals will sometimes inject the trees with concentrated glyphosate, but I'd leave that to them.

Sow Thistle

Sonchus arvensis

Common Names: Perennial sow thistle, corn sow thistle, hare's lettuce.

Family: Asteraceae
Genus: *Sonchus*

Looks Like: a lanky, not very lovely perennial. A very closely related variety, *Sonchus oleraceus*, is an annual. Sow thistle looks like the gangly lovechild of a dandelion and a pair of chopsticks. Rising from a taproot, the leaves are coarsely toothed and up to a foot long, light to lime green, in an alternate arrangement. The stem is occasionally branched, smooth, and hollow between the joints. The flowers show up in midsummer; little yellow daisy-like creations about an inch across and gathered in clusters at the top of the stalks, which can be anywhere from 4 to 10 feet tall. The flowers dry into white parachutes of fluff, and the seeds blithely drift away to seed somewhere new. When any of the sow thistles are broken, non-toxic and milky sap dribbles out.

Comes From: Europe, somewhere or other. Nobody can say these days, since it's so ubiquitous now.

Likes To Live: oh, anywhere it can find half an inch of land, some moisture and a good bit of light. It likes neutral soils and loam, but it'll take most anything.

States Present: Most of the Midwest and more temperate states.

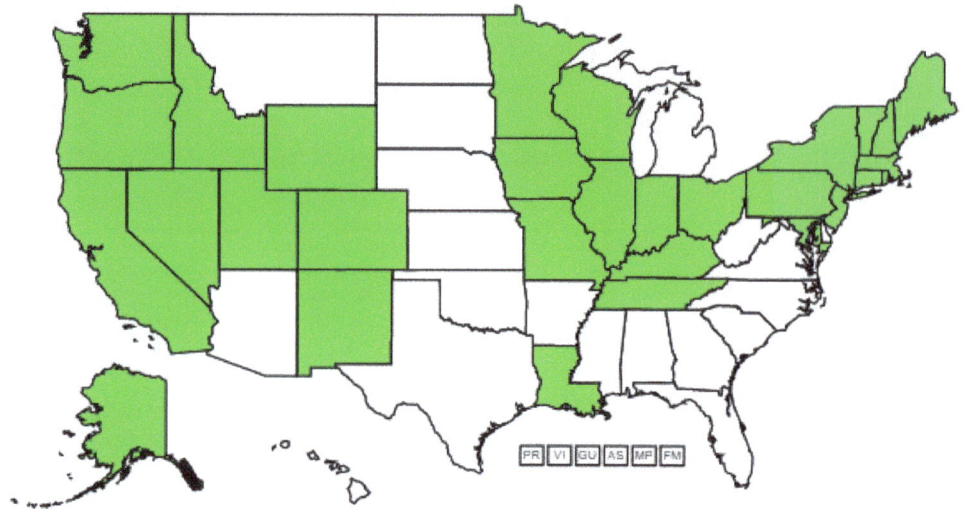

States Impacted: All the ones where it grows.

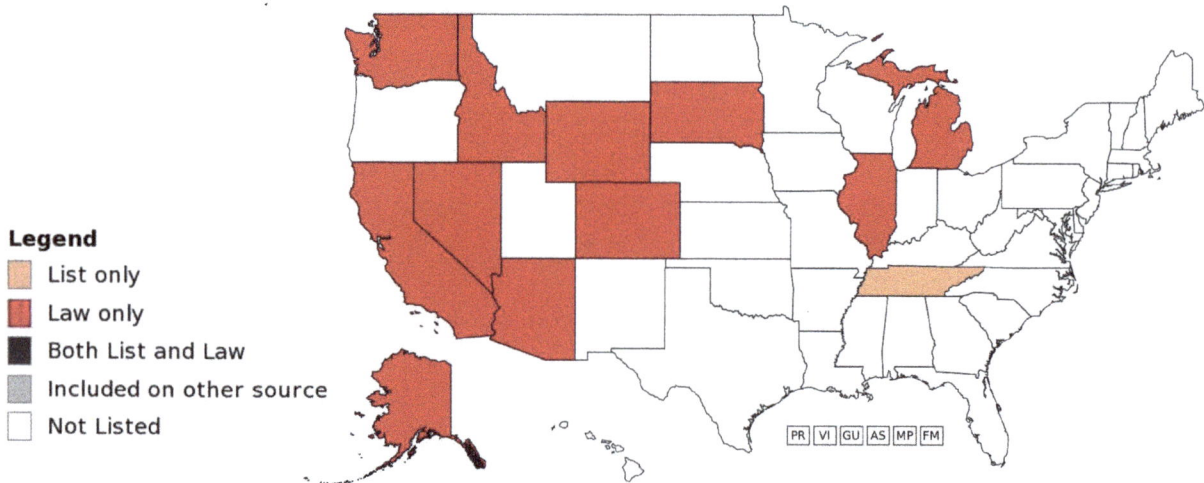

Legend

- ☐ List only
- ☐ Law only
- ■ Both List and Law
- ☐ Included on other source
- ☐ Not Listed

Succeeds By: seeding like anything. Those sleepy, gentle-looking sow thistle seeds floating on the wind are drifting there with anywhere from 4,000 to 13,000 siblings. And they're all ready to get growing.

Causes Trouble By: competing with economically valuable crops. It's a problem in several crop types, where it causes reduced crop yields, increased cultivation and herbicide expenses and land depreciation. Studies from the University of Michigan show that densities of 17 and 32 sow thistle shoots per yard reduced wheat yield by 15 and 45%, respectively. Heavy infestations— around 83 plants per yard— have been shown to reduce soybean and dry bean yields by 49%. In dry years, these effects are exacerbated, and 80% reduction in the yields of both crops have been recorded.

Sow thistle was first reported in 1814, in Pennsylvania. The first sample was collected in Maine in 1894. Hailing from Europe, it seems to be one of those plants that meandered over here with no particular aim in mind, hitching a ride along with crop seeds. A bit of a vagabond, it wanderers about aimlessly, doing no particular good for anyone. The genus Sonchus, the old Greek name for 'hollow', was applied to the hollow stems. The species name, *arvensis*, means 'of planted fields', which it certainly is. Like a drifter, the plant doesn't do terrible harm... but it doesn't do much good either.

If You Love It: ...reconsider, perhaps? It's technically edible, but I'd reach first for its close relations, chicory and dandelion. They won't stab your hands nearly as much.

If You Loathe It: then at least it's easy to remove.

For the Home

- **Don't let it get started:** Keep an eye on fill dirt, hay and seed from outside your area. Make sure what you bring into your garden is clean.
- **Yank it:** Wearing gloves, just pull the plant out of the ground. It comes out quite easily. Throw it in the compost if it hasn't started to seed. If it has, trash it.

- **Catch it young:** The seedlings and rosettes are easy to kill; just pull them or use a fire wand on them.
- **Mulch it:** The seeds won't sprout if they can't see light, so that's a very easy fix.
- **Compete with it:** A good ground cover will be stronger than the opportunistic thistle over time. Amend poor soil and irrigate, and other plants should take off.
- **Watch it:.** Keep after young sow thistle plants for a few years, until the seed bank in your soil is exhausted.

For Larger Spaces:

- **Mow it:** Sow thistle isn't hard to control with a mower.
- **Graze it:** Cattle, sheep and goats will happily eat it, and it's good for them. Horses don't like it.
- **Compete with it:** Plant ground covers or grasses that beat the thistle. Amend poor soil if that's the problem. Sow thistle is a symptom of degraded land, not the cause. Fix the problem, and the land will heal.

The Chemical Option: is a bit of overkill in this case. 2,4-D is effective, but a fire wand isn't as dangerous. If you are going to spray sow thistle, be smart: cut the top off and spray the bottom to kill the roots. Don't spray an entire ten foot tall plant. It's a waste.

Tree Of Heaven

Ailanthus altissima

Common Names: ailanthus, copal tree, stinking shumac, stinking sumac, false varnish tree, Chinese sumac, paradise-tree, stinkweed, stink tree.

Family: Simaroubaceae

Genus: *Ailanthus*

Looks Like: a rapidly growing deciduous tree up to 80 feet tall and 6 feet around. It has large, heart-shaped leaf scars on the twigs. The bark is smooth and light grey, developing fissures as it grows older. The leaves are pinnately compound and 1 to 4 feet long, with up to 41 leaflets. It's a bit like sumac, but the notched base on each leaflet gives it away. Oh, and the smell too. The whole thing smells like peanut butter gone bad.

The tree is dioecious (separate male and female plants). Flowering happens in early summer, when large clusters of small yellow flowers on 20 inch long stems rise above the leaves. They ripen into tan to reddish, single winged keys on the female trees, with one seed in the middle and can be spread by wind or water.

Comes From: The Beijing area of China.

Likes To Live: just about anywhere. This invasive tree species thrives in degraded soils, and can even grow in cement cracks. It isn't very shade tolerant, but it easily scrambles along forest edges, causing habitat damage. And it's quite happy to take over vacant lots, construction sites, and anywhere where humans have wrecked the existing plant community.

States Present: 46 out of the 50. Only the Dakotas, Montana, and Wyoming avoid it.

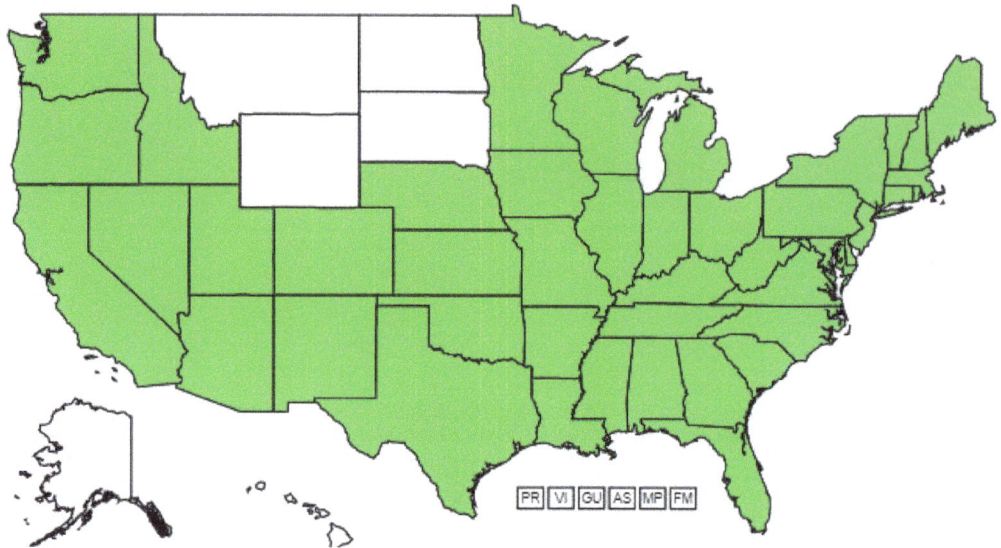

States Impacted: Every one where it grows, more or less. It's a particular problem in New York and the East Coast.

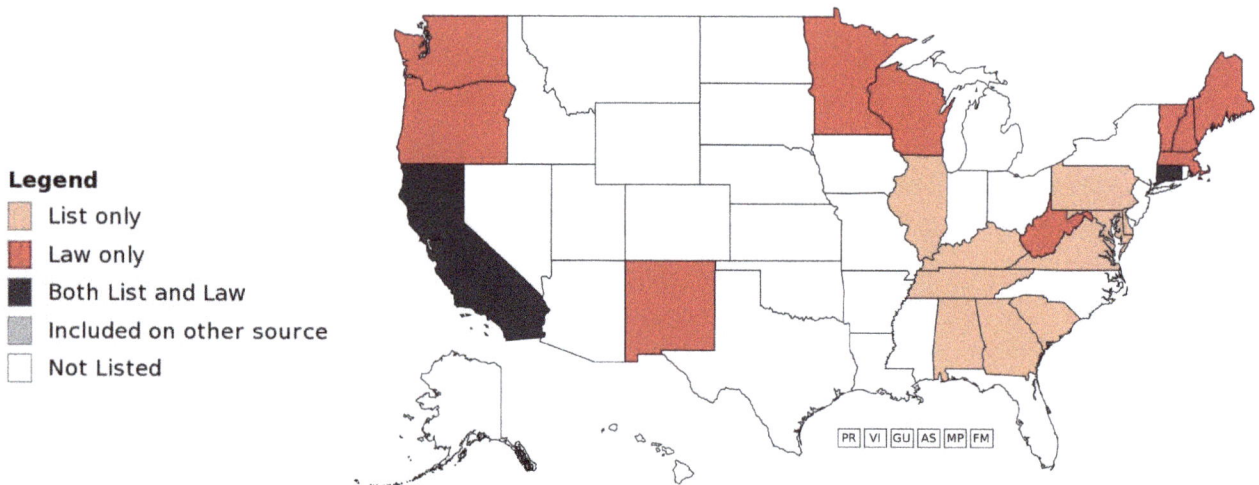

Legend
- List only
- Law only
- Both List and Law
- Included on other source
- Not Listed

Succeeds By: being incredibly good at both seeding and clonal reproduction (sending up shoots from the roots). A single female tree can make up to 325,000 seeds a year. That's 5 million seeds in her lifetime. Both genders reproduce clonally from roots that feel sunlight on them. Established trees continually spread by sending up root suckers; these can emerge as far as 50 feet from the parent tree. An injured tree of heaven may send up dozens of root sprouts, and those sprouts can produce seed within 2 years.

Causes Trouble By: forming dense, clonal thickets that can rapidly invade fields and meadows, as well as native and managed forests. Since nothing wants to eat it, it beats native plants to maturity, steals their light and crowds them out. Which means the animals that normally feed on native plants are out of luck, too.

Oh, and it poisons other plants; the root system releases a chemical known as ailanthone, which suppresses the growth of other plant species. And it is very hard to kill. If a tree is cut down, it aggressively resprouts from the roots, making controlling it difficult. Its aggressive root system can cause damage to pavement, sewers and building foundations; by the time you notice the heaving and the cracks, it's already too late.

The plant also acts as a host for the spotted lanternfly, an invasive insect who also came over from China. These insects seek out the tree of heaven as a place to lay their eggs. Mature lantern flies feed on and damage a lot of native and fruit-bearing trees. Putting all these details together, the economic cost of controlling the tree and its effects is a little scary.

The tree was introduced to the U.S. in 1785 by William Hamilton of England. Hamilton got that first *Ailanthus* tree from England, courtesy of Kew Gardens. Hamilton planted it at his Philadelphia estate, known as The Woodlands. He even gave a tree to a member of the Bartram family as a wedding gift.

The earliest known seed production was found at the Philadelphia Herbarium and the Academy of Natural Sciences in the early 1800s—likely with seeds from Bartram's Garden, still the oldest known botanic garden in the nation. In Philadelphia, tree of heaven sparked the interest of amateur and professional horticulturists alike as a lovely ornamental tree for the gardens of larger home and farm landscapes. The seeds and seedlings were sold and traded between gardeners to shade their own landscapes.

The second introduction happened in 1820, at Long Island. At this time the seeds and seedlings were well in with the horticultural community. Horticultural societies reported that the seeds of the tree were distributed by the U.S. Department of Agriculture, as a free shade tree.

The third U.S. introduction of tree of heaven took place in California, during the mid-1800s. The Chinese work force brought in during the Gold Rush brought seedlings; for them, it was a cultural reminder of home and a useful living pharmacy.

The tree was valued, at first, as a unique, fast-growing ornamental shade tree for cities. It has the ability to tolerate saline soils, air pollution, a range of soil pH and drought, making it a perfect tree for city habitats. It was widely planted from New York City to Washington, D.C. But by the early 1900s, the tree began losing its popularity. The tree in question couldn't care less. It happily grew away anywhere it could force itself up between a couple paving stones.

It really began to spread during the 1980s, when a massive gypsy moth infestation threatened huge sections of forest. The moths defoliated hardwoods, which were then logged by salvaging operations. Whole swaths of forest lands were left open to plant invasion. Tree of heaven jumped on the chance and scrambled into the Eastern forests, where it put down rhizomes which spread horizontally underground, crowding out native species. Tree of heaven made itself at home, and getting it to leave is not going to be easy.

If You Love It: It does have a lot of beauty. And in the difficult conditions of a city, it can be a blessing to see something green in a barren place. But please, do not encourage this tree to spread. It's doing quite well enough on its own. That said, if some is already around, put it to use. The wood is satiny, yellowish-white, great for wood-working. Wear a face-mask though; the leaves and fresh sap stink.

If You Loathe It: Let's talk about control and, just possibly, eradication.

For the Home

• Hand-pulling young seedlings is effective when you can get the whole root. Look out; small root fragments can generate new shoots.
• Seedlings can be easily confused with root suckers, which are nearly impossible to pull by hand.
• Plants that are too difficult to pull or dig should be cut down to a 1 inch high stump and a glyphosate-based herbicide applied to the freshly cut stumps within five minutes, using a paint brush or sponge applicator. See the end of the book for safety details on this process! Stump application is very effective during July, August, and up to mid-September.
• If it re-sprouts, cut and poison it again. You may need to do this several times over the course of two summers before the root system finally gives up the ghost. Sadly, almost nothing else is really effective for killing this devil with an angel's name.

For Larger Spaces:

• **Take advantage of shade:** Where the tree of heaven is shaded, sprouts and suckers will still emerge, but they'll gradually slow down and die unless they can reach sunlight. Try planting large shade trees to outcompete it, or put a tarp over an area like a small lot, to block the light.
• **Cut them early and often:** Young tree of heaven trees have not had as much time to develop the extensive root system that will promote vigorous sprouting and suckering. A mower or a weedhog is very effective in the first few years. Sprouting and suckering will still occur, but will be easier to manage than if you allow the tree to get large. Run them over 4 times in each growing season.
• **Grazing/Browsing:** Cattle, sheep, goats, and deer will all eat tree of heaven under the right circumstances, though it does not seem to be a favorite food for any of them. It has been

185

reported that goats will eat both the leaves and bark at certain times of year. Where livestock grazing pressure is heavy, it may be controlled. But don't bet on it.

Keep after tree of heaven for about three years, and you'll beat it eventually. Just stay on top of it, or it'll go way over your head. Literally.

Tumbleweed

Salsola tragus

Common Names: Russian thistle, prickly Russian thistle, kochia, wind witch.

Family: Chenopodiaceae
Genus: *Salsola*

Looks Like: an annual herb that can grow to 4 feet tall. The branches begin life in a whorl, branching straight from the central leader. The leaves are alternate and narrow; the upper leaves have a sharp point at the apex and are 2 inches long. Flowering occurs from midsummer to fall; small, pinkish or greenish flowers develop from the leaf axils. Each flower is subtended by 3 spine-tipped leaves.

As the plant matures in October, the older leaves become short and stiff, with a sharp-pointed tip. The overall shape of the plant becomes oval to round as the branches dry and curl inwards. At maturity, the plant pales down to the color of straw. It can be anywhere from 18 inches to 6 feet wide. After the plant dries, the base of the stem becomes brittle, and the plant breaks off at soil level in early winter. These round, spiny plants can roll along with the wind for miles, spreading their seeds. Until they run into something, of course.

Comes From: Russia.

Likes To Live: in vacant lots, abandoned gardens and agricultural fields, roadsides, fencelines; really, any open site with loosened soil. Tumbleweed is extremely drought tolerant. The taproot can extend several feet into the soil. It's especially well adapted to desert environments; it can tolerate alkaline soil conditions and is very competitive when water is scarce, when soils are disturbed, or when competing plants are stressed by overgrazing.

189

States Present: All the western and northern states.

States Impacted: All the ones it grows in.

Legend
- List only
- Law only
- Both List and Law
- Included on other source
- Not Listed

Succeeds By: seeding and, innovatively, rolling! The seed is spread when mature plants detach at the base and are blown along by the wind. A large Russian thistle plant may produce more than 200,000 seeds. Being spiky, the dead adult acts as a rolling nursery for the seeds, protecting them from predation as it spreads them in their new homes. In spring you can trace the paths of tumbleweeds across plowed fields by the green trails of germinating Russian thistle seedlings.

Causes Trouble By: Doing all kinds of problematic things. In agricultural areas, Russian thistle can reduce yield and quality of a number of crops, particularly alfalfa and small grains. It depletes the soil moisture, interferes with tillage operations, and serves as a shelter or food source to many problematic insects and crop diseases. It's an alternate host for curly top, a disease that affects potatoes, beans, and a few other crops. It can also threaten native plant ecosystems by smothering plants in its rolling husks and robbing the native plants of moisture with its seedlings.

And then it gets a bit Western Gothic. Large plants can reduce highway safety by obstructing views along right-of-ways and causing drivers to swerve their cars in an attempt to avoid colliding with huge accretions of the windblown plants, bowling across the road like some ghost of the buffalo. In many areas, tumbleweeds pile up along tree rows and fence lines. If they catch fire, they burn very hot and very high. This means that large piles of them absolutely must be cleaned. Since they're large, spiky, and awkward, this takes an awful lot of labor.

And it's not good for people to mess with tumbleweeds, either. A lot of people are sensitive to Russian thistle, and a little scratch from the plant may result in itching or reddened patches of skin. The windblown pollen of Russian thistle can cause an allergic reaction in people during summer.

When winds are high and strong, the tumbleweeds make travel truly dangerous on highways; they can completely fill roads and bury cars. No, this isn't a cowboy's tall tale. It's in the news. In 2018, Victorville, California became a news item when it was all but buried by tumbleweeds after a windstorm. And as the climate warms, the tumbleweed problem gets worse. Russian thistle is a major problem along the California aqueduct, where it can fill up and block the waterways. Want the really scary bit? It has been reported that prairie wildfires spread when ignited balls of burning Russian thistle blow through grasslands. Now that's a tale for the campfire.

When you see a tumbleweed roll, you may imagine the sound of a menacing whistle. Tumbleweeds are the spirit of the American frontier in our cultural imagination: nomadic, lawless, rough and dusty. A little strange and a little magical.

That's what we picture when we think of the Wild West. But that tells us much more about ourselves than it does about the plant, because it isn't a natural part of America at all. It comes from the other side of the world.

In 1873, a group of Russian immigrants moved into Scotland, South Dakota, to start new American lives. With them they brought everything they needed to make a fresh start: home goods, good work ethic, and sacks of seed from the old country. Among them were sacks of flax, which were planted in their season. Soon, a new plant was springing up beside the flaxseed. These were the first tumbleweeds in America. Because it thrives on little water and treats land stripped of native species like its own personal planting bed, it took hold in the vast agricultural fields and overgrazed rangelands of the arid West like nobody's business. Just 15 years later, the tumbleweed (also called the Russian thistle) had rolled its way from Canada and California. And we're still dealing with that mistake to this day.

The story of the tumbleweed tells us about the power of stories and myth-making. It's woven into all our stories of the American West, tied into our concept of what this lonesome prairie is. But the presence of the tumbleweed tells us that this is a story that centers the settler, the immigrant and the stranger. Where is the story of the Native West, the one where people worked *with* a dry land, and didn't try to force it to become something else?

That story is beginning to be remembered, spread, and known better: a place where people grew amaranth and drought-resistant beans, melons and cactus fruit. A West where the people knew where pinyon seeds were gathered, and worked with the land instead of treating it as their servant or their enemy. This way of farming is being remembered. Places are being founded like Ndée Bikíyaa—The People's Farm, on the White Mountain Apache Reservation. Ndée Bikíyaa works to reconnect the community to its food, traditional lifestyles and, ultimately, its balanced and healthy state of being.

The European-American stories of the West are like tumbleweeds, rolling over and thoughtlessly burying what is below. But like the agave and the saguaros, the truths of the West will patiently wait until the tumbleweeds blow away, and they are revealed again.

If You Love It: Sing about the tumblin' tumbleweed and indulge in a little nostalgia, but don't let the plant seed.

If You Loathe It: Eradication measures are starting to be nailed down.

For the Home

- **Dig:** Digging up the seedlings when they're young can prevent seed production.
- **Take care:** Try not to loosen the soil, because loose soil is necessary for tumbleweed seeds to sprout.
- **Burn it:** Burning is sometimes used to destroy accumulated Russian thistle plants. Just be smart and careful about this. Do it well away from structures, on a calm, windless day when the fire damage is low. Wet down the ground that is going to be your fire area. Don't make problems bigger.
- **Outcompete it:** Planting competitive, desirable native grasses and groundcover species can be a great way to stop Russian thistle establishment. Russian thistle competes poorly in situations with firm, regularly irrigated soil, and it is rarely a problem in managed gardens, turfgrass, or landscapes. Herbicides aren't needed in home gardens and landscapes for Russian thistle control.

For Larger Areas: What's good for the small space is good for the large space. Grazing animals on the young plants can keep them from seeding. If that doesn't do the trick, there are herbicides that will control Russian thistle in agricultural crops. Aim treatments at controlling the immature plants to prevent them from producing seeds; that's the key. The selection of an appropriate herbicide depends on the site and the crop.

Pre-emergent Herbicides: Pre-emergent herbicides are applied to the soil before the weed seed germinates and usually incorporated into the soil with irrigation or rainfall. They're one of the best tools against the incredible seed bank of tumbleweed.

Be aware that herbicide-resistant types of Russian thistle have evolved in only a couple of years following treatment with chlorsulfuron (Telar) or sulfometuron (Oust). Avoid repeated use of a single herbicide or of herbicides that have the same mode of action to prevent the evolution of herbicide-resistant populations.

Post-emergent Herbicides:

Post-emergent herbicides are applied to growing plants, but the timing is absolutely critical. For best results, these herbicides must be applied while the weed is in its early growth stages, preferably the early seedling stage, before it becomes hardened and starts producing its spiny branches. The spiny stage of Russian thistle laughs at herbicide.

Two-spotted Spurge

Euphorbia maculata
(currently being updated to Chamaesyce maculata)

Common names: spotted sandmat, large spurge, spotted spurge, prostrate spurge, milky purslane, milk-purslane, eye bane.

Family: Euphorbiaceae
Genus: *Euphorbia*

Looks Like: a warm-season annual that grows close to the ground, forming a thick mat. Its dark green leaves, growing in opposite pairs, are 1/8 to 1/2 inch long and about 1/8 inch wide. A red spot usually marks the center of each leaf. The short stems have a separate stipule—or little scalelike appendage—at their base, although you may need a hand lens to see them. Broken stems and branches bleed a milky, caustic sap. Don't get it on your skin, and whatever you do, keep it away from your eyes!

Spotted spurge produces tiny, pinkish flowers that consist only of stamens and pistils, grouped in small, flowerlike cups called cyathia in the leaf axils, the area where the leaf joins the stem. The fruit is a 3-celled seed capsule that is 1/16 inch or less. Each cell contains one seed that is about 1/25 inch long. The plant's central taproot system extends more than 24 inches into the soil.

Comes From: the Eastern Seaboard of North America. It started as a scrubby seashore plant, but now it's everywhere.

Likes To Live: anywhere other plants struggle. It thrives in poor, drought-stressed turf, thrives in hot, dry weather, and grows well in thin soils. It often is found on closely mowed sites and on edges of lawns next to curbs, driveways, and sidewalks. It is low growing, but can grow over short turf and spread out and form a mat, choking out the grass.

States Present: Everywhere but Alaska.

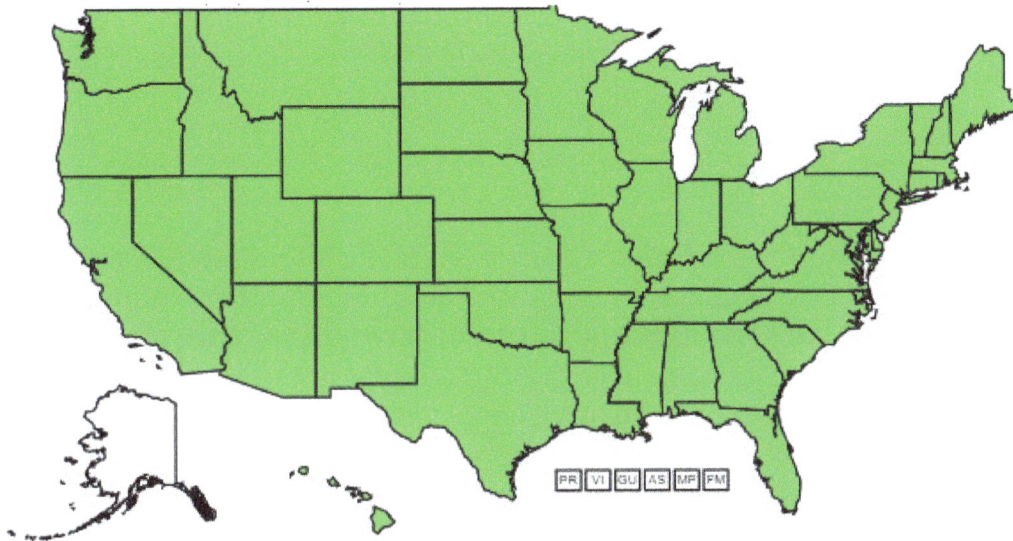

States Impacted: All the ones where it grows.

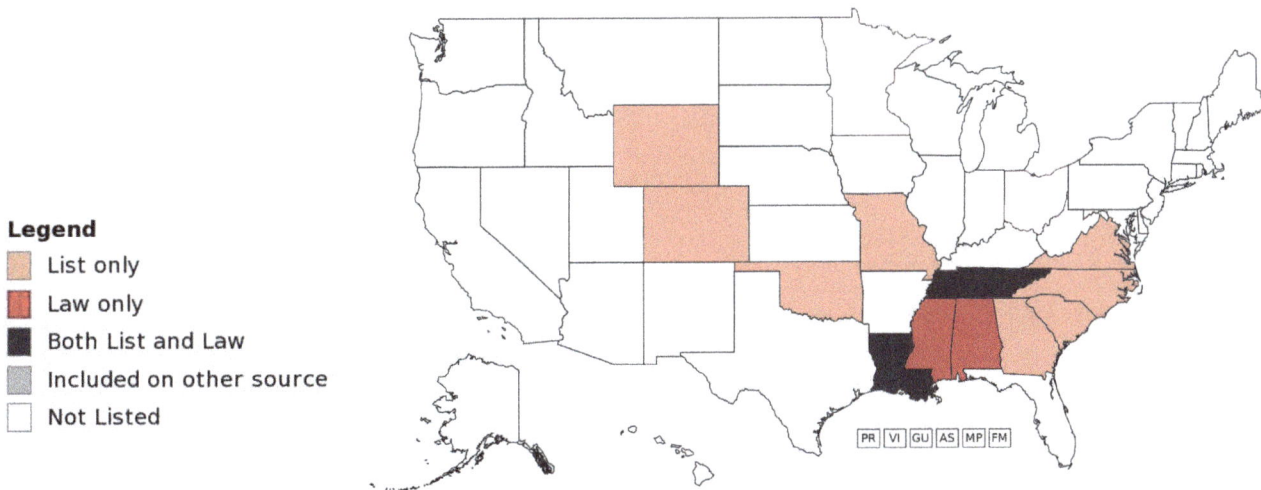

Legend
- List only
- Law only
- Both List and Law
- Included on other source
- Not Listed

199

Succeeds By: seeding like nobody's business. Each minute brown seed is hairy, with longitudinal ridges or wrinkles. The seeds stick to surfaces when they get wet, including animal fur, human's shoes, and anything else the slightest bit damp. Plants can begin producing flowers and seeds just over a month after germination. Thousands of seeds are produced by every plant, which can germinate immediately, the following spring, or years later. Thousands of seeds may be produced by a single plant. Buried seeds can be viable for over 50 years!

Causes Trouble By: stealing resources from other plants when things are toughest, and making lawns look scruffy. As weeds go, prostrate spurge is more of an annoyance than a real threat to crops. In really bad infestations, it can grow over and smother other plants when they're young, but that means the desirable plants were already not doing great. Spotted spurge is the plant equivalent of the common cold; never a real danger, but always a nuisance.

It can establish itself in horticultural, agricultural, and landscaped areas. It overgrows sparse turf areas and low-growing ground covers, invades open areas in gardens and landscapes, and even sidewalk cracks! It provides a habitat for undesirable insects in citrus groves, serves as an intermediate host for fungal diseases of cultivated crops, and attracts ants with its seed.

But it does pose one serious threat: it's toxic. Spotted spurge can kill sheep grazing in pastures where it is the main weed. Sheep that consumed as little as 0.62% of their body weight of this plant have died within a few hours. It's also dangerous to have around children and klutzy people. The milky sap in spurge causes dermatitis, and the fresh plants must be handled with care. If the sap gets on the skin, it must be washed away immediately. In the eyes, it's extremely caustic and will require a trip to urgent care.

A passing reference of Pliny's copied over by Culpeper says the sap has been used against warts, but even these ancient books urge caution. It makes sense, as the plant is caustic… but I wouldn't use it.

This is another one that we have to glare at our ancestors for. Robert Prince, of Flushing, New York, opened the first nursery in the United States in 1737. He named it the Linnaean Botanic

Garden and Nurseries. Well-supplied with exotic plants from around the world by sea captains he'd worked with in the previous decades, Prince and other nursery pioneers made it possible for homeowners to acquire all kinds of plants.

These plants, gathered from around the world and grown in New York state, were packed tightly and grown fast, to be sold all up and down the Eastern seaboard and beyond. The nursery garden was laid out with dense plantings in a rectilinear plan that permitted, as its seed catalog describes it, "plenty of stock in a little space."

This means that thousands of plants were packed tightly and being irrigated with more or less efficiency. Into this variable new environment, a little weed of the New York seasides and marshes crept in and started to flourish. When the plants of Prince Nursery were traded to botanic gardens, sold to the public, and shipped, the seeds of spotted spurge went with them.

Prince Nursery was managed by four generations of the Prince family, until it eventually closed around 1865. The second-generation owner of the nursery was William Prince, who is considered to be the founding father of the nursery industry. He turned his father's nursery into a true business. Prince did a great deal to introduce new plants across the country. Intentional and otherwise.

If You Love It: ...I really have to wonder why.

If You Loathe It: There's a lot that can be done to get rid of the little beast.

For the Home

● **Care:** First, as always, cultivate your property in a way that doesn't encourage spurge to grow. Infestations of spotted spurge in newly planted lawns can be cut down by seeding your turf in late summer or fall, out of the spurge's germination period. In mature lawns, improve the turf density through fertilization, regular appropriate mowing, and using turfgrasses adapted to your site conditions. Remember, run your mower blades high! Healthy turf easily outcompetes spurge.

- **Mulch:** The seeds must see the light in order to germinate. A good layer of quality mulch prevents this.
- **Watch and Pull:** Regularly check the infested areas, and pull the new plants before they produce seed. Be sure to get the roots. Wear gloves when you pull the plant or use a dandelion-fork/hoe; the sap can be a skin irritant. Mowing isn't going to work with spurge; it grows too low. In my experience you barely have to pull it; scuff it out of the ground with the toe of your boot and it dies. It really depends on seeding to spread, and its roots aren't particularly tough.
- **Solarization:** Before planting an area with turf or ornamentals, you might want to follow the management method known as soil solarization. See the index of control methods in the back of the book for details.
- When planting new, container-grown ornamentals and ornamental beds, be sure to use sterilized planting mix. When purchasing plants for ornamental beds, check the plants for spurge seedlings.

For Larger Spaces: Tilling is not effective; you're just prepping a seed bed for a bad infestation of spurge. Animals can't eat it. Better get out the chemicals.

The Chemical Option: Pre-emergent herbicides can be a big help to prevent spotted spurge outbreaks if you apply them in late winter, before the weed seeds germinate. This weed can be controlled with various post-emergent herbicides.

Wild Mustard

Sinapis arvensis L., Brassica campestris, Brassica nigra, Brassica oleracea

Common Names: Black mustard, brown mustard, rocket, poor man's cabbage, yellow cress.

Family: Brassicaceae
Genus: *Sinapis*

Looks Like: a perennial herb, annual in colder environments. Upright, sparingly branched stems have blade-like leaves and can reach 3 to 6 feet high. The lower leaves are stalked and fairly large, around 18 inches long, with irregular wavy margins. (please note that I'm covering several almost identical species here, but each has its own quirks.) Clusters of tiny, yellow flowers with four petals bloom early in spring. There are around 20 to 40 individual flowers. The flowers ripen into short-beaked siliqua (fruit divided into two cells by a thin partition) around 1/4 inch long, with a seed or two in each.

Comes From: the Mediterranean region and Southwestern Europe, as far north as Southern England. It is found growing wild on seaside cliffs and stony land on the coast.

Likes To Live: in open land: cultivated fields, gardens, pastures, riverbanks, roadsides and disturbed places.

States Present: Every one.

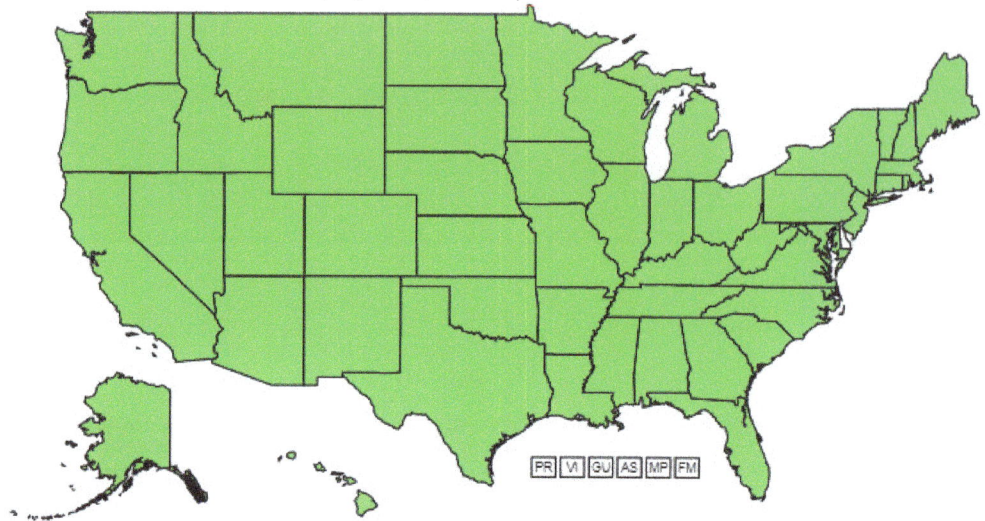

States Impacted: All agricultural states have a mild problem with it.

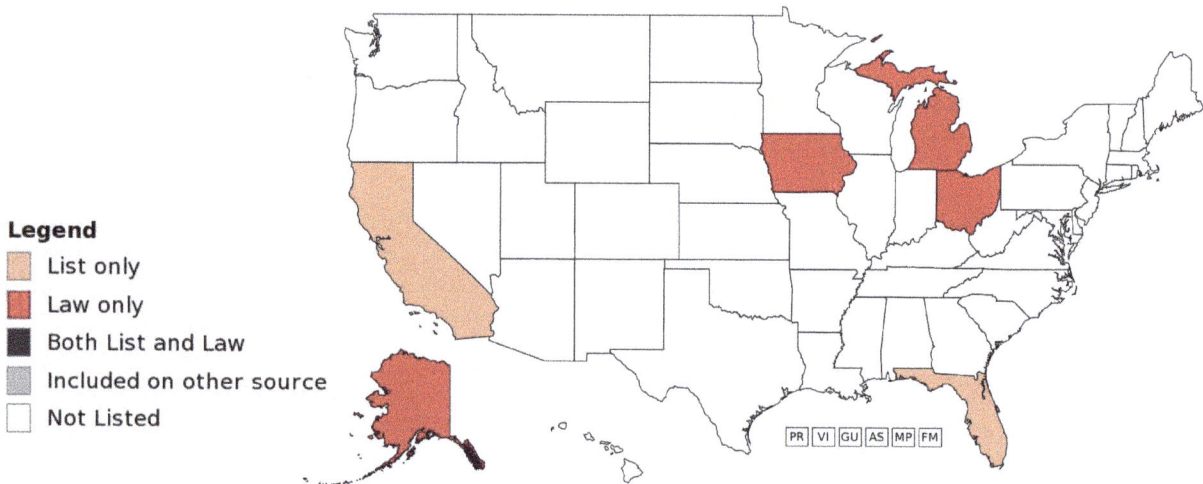

Legend
- List only
- Law only
- Both List and Law
- Included on other source
- Not Listed

206

Succeeds By: Seeding. Each plant can produce up to 5,000 seeds a season, allowing it to spread rapidly.

Causes Trouble By: being a serious weed of cultivated land and generally far too enthusiastic in its seeding. It is responsible for drops in crop yields, crop value losses, and for costly chemical and cultural controls. The presence of wild mustard seeds in harvested canola seed results in a serious loss of quality in canola oil and meal. Since wild mustard seed is similar in size and shape to canola seed, it is impossible to separate the seeds using conventional methods.

Wild mustard is an alternative host for a number of pests, including insects, nematodes, fungi, viruses and bacteria that attack cultivated crops, especially members of the Brassicaceae family. Important crops in the Brassicaceae include broccoli, cauliflower, Brussel sprouts, and cabbage. Since all these plants were bred out of wild mustard originally, it's really no surprise.

In spring cereals, dense wild mustard infestations reduce yields by as much as 53% in wheat, 63% in oats, and 69% in barley. It causes trouble in natural areas, too. Wild mustard is loathed among native plant activists. It smothers native plants and flowers, transforming the landscape of the United States. When it dries in the summer, it fuels wildfires in the West. Not good.

Stories are many and varied about how wild mustard arrived in the United States. Two romantic tales relate that either Gaspar de Portola brought the seeds from Spain and planted them to mark his trail as he led his expedition north from Mexico in 1769 to found the missions in California, or that they hitched rides on the livestock Juan Batista Anza brought to Southern California with the first colonists in 1775. A more prosaic report suggests that wild mustard seed arrived in Northern California in the 1800s along with Russian settlers, who carried it in sacks of wheat they imported to their new home.

Personally, I wouldn't be surprised if it arrived in a number of places for many reasons; planted in gardens in New England as a spice-herb, accidentally mixed into Russian seed, and used as a pretty mile-marker too. That's because it's a plant that's always been with European cultures. Mustard has been an important part of the human diet since at least the Bronze Age.

The sharp flavor of the seeds offered one of the few opportunities for Europeans to spice up their foods. In modern times, cultivated mustard is second only to salt and pepper in our favorite condiments. It's possible that you had black, brown, or white mustard seasoning your food in a recent meal. And chances are also good that several other parts of your dinner might have come from the same family. High in vitamin C, the brassicas have been bred into all sorts of vegetables. Kale, rutabaga, seakale, turnip, radish, watercress, horseradish, wasabi, and kohlrabi are all bred from this little yellow-flowered plant. A number of common dinner vegetables including cabbage, broccoli, cauliflower, Brussels sprouts, and collard greens, are all cultivars of just one species: Brassica oleracea. So, when you drive by swathes of bright yellow flowers in spring or find one gangly little yellow-flowered weed in your garden, give a nod to the bountiful and generous mustard family.

If You Love It: The forager's golden rule, 'harvest only one third of any stand of plants', can be ignored when harvesting wild mustard. Harvest the whole plant, and rip out a few more while you're at it. Eating wild mustard helps reduce this invasive species and gives your local plants a fighting chance! As a food, it's very nutritious. It's 25% protein and rich in calcium, phosphorus, magnesium and vitamin B as well as vitamins A and C.
The flavor is hot and tangy, and can be eaten as a raw green to spice up salads or treated like spinach. It also can be boiled or steamed, but some of the spiciness is lost.

The seeds are great pickled, dried, crushed, or mixed with vinegar and other flavors for condiments. White seeds are best for "ballpark yellow" mustard, while the brown variety offers the best greens for sautéing or stewing. To get your seeds, clip the ripe seed pods before they release the seeds. Dry them in an open tray, then put on rubber gloves and rub them between your hands to get the seeds out. Do be careful to wear gloves; the oils in mustard can burn the skin when concentrated. In fact, you'll read about mustard blisters in old herbal books, suggested for drawing out ill humors and ending pneumonia… but I really don't recommend trying it.

The fresh flowers offer beautiful color and both a sweet and spicy note to dishes.

If You Loathe It: It's pretty easy to control.

For the Home

- **Pull it:** They come out of the ground very easily when young, and can't readily resprout from root fragments.
- **Hoe it:** For areas of young seedlings, hoe them all up by scraping a shuffle hoe over the ground.
- **Mulch it:** A good covering of mulch will prevent the seeds from germinating.
- **Eat it!** Pull the plants up wherever you find them and use them in the kitchen. Just make sure they don't have pesticide residue on them.

For Larger Spaces

- **Mow it:** Preventing the seeds from germinating is key. Mow the plants throughout the growing season to keep it from setting seed.
- **Till and Cover it:** Tilling the property and putting in a strong cover crop is a great way of outcompeting the weed.
- **Graze on it:** The plants are very nutritious for animals.

The Chemical Option: A spraying of 2,4-D is very effective.

Wild Parsnip

Pastinaca sativa

Common Names: cow parsnip, bird's nest, common parsnip, poison parsnip, hart's eye, boilwort, wild mustard, devil's mustard.

Family: Apiaceae
Genus: *Pastinaca*

Looks Like: a tall biennial or perennial herb that looks and smells similar to cultivated parsnip. The leaves are alternate, compound and branched, with jagged teeth. Stems are erect, branched, hollow, and slightly grooved. Leaflets are yellowish-green, shiny, oblong, coarsely-toothed, and diamond-shaped. The plant can grow up to 6 feet tall in an average year. Flowering occurs from May to June, when hundreds of yellow flowers develop. Flowers are arranged in an umbel; for those familiar with Queen Anne's lace, this looks like a yellow version. The fruits are dry, smooth, slightly winged and flattened on the back. The fruits each contain two seeds, which disperse in the fall.

Comes From: Eurasia.

Likes To Live: anywhere with sun. It can survive in a wide range of environmental settings, from dry soils to wet meadows. It is commonly found growing along roadsides, in pastures, and in abandoned fields, along highways and railroads, or any place where the soil has been disturbed and native vegetation has yet to re-establish.

States Present: Everywhere but the Deep South.

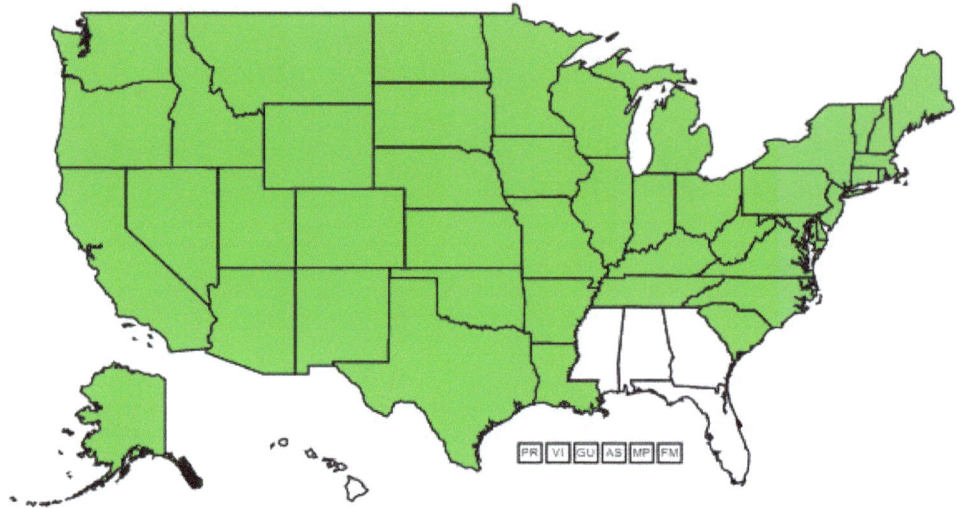

States Impacted: The Midwestern and Eastern states, where rainfall is plentiful.

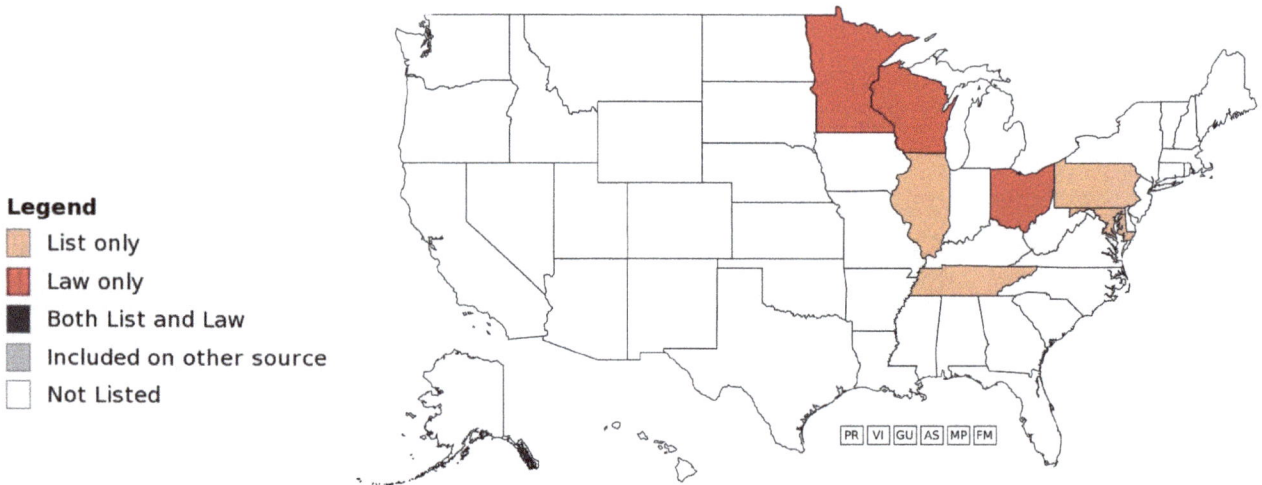

Legend
- List only
- Law only
- Both List and Law
- Included on other source
- Not Listed

Succeeds By: seeding prolifically and being adaptable as a chameleon.

Causes Trouble By: making itself really, really unpleasant. Like giant hogweed, it's in the carrot family, and it's nearly as toxic as its taller relation. Both the wild and cultivated forms of parsnip contain toxic compounds, called furanocoumarins. These compounds can cause serious rashes, burns, or blisters to skin exposed to the sap and then sunlight. The chemical content is higher in the wild species, and it is very nasty. Being exposed to the sap causes human and animal skin to become photosensitive; exposure to sunlight once its chemicals have been absorbed causes severe blistering. Very painful, very irritating blistering. I speak from experience here. The blisters often lead to scarring. Sap in the eyes can cause blindness. The plant poses a risk to agricultural workers, those involved with vegetation control, and to people exposed to the plant in the wild.

It outcompetes native vegetation, particularly crowding out lower-growing plants. The tall canopies shade out other plants and rob them of sunlight. It can also have an impact on pollinators, as honeybees do not visit the plant and it may displace other, more pollinator-friendly plants. All around, it's a problem.

I can only stare wide-eyed at the history of this plant, even now. Was it created as a bio-weapon in a lab? No. Was it a terribly mistaken introduction that made our ancestors rue the day they put down the seeds? No.

It was *planted*. As a *food crop*. A root crop, to be specific. It is only a few genes away from the domestic parsnip that is used in stews, beer, and various dishes to this day. For fans of domestic parsnips (*Pastinaca sativa*) you'll be able to tell this foe from your friend by its size. While domestic parsnips can be up to 4 feet tall, the titanic wild parsnip is much larger and has much bigger, deeply incised, rounded leaves.

I grew up in rural upper Wisconsin, where wild parsnip grows rampant in untended fields, works its way into horse pastures, and is much hated. I still have a few scars around one ankle because of an encounter with wild parsnip when I was nine, a remnant of the blisters that resulted. The first time I learned the history of this plant, I burst out laughing incredulously. Who in the world would want to get anywhere near a plant that blisters your skin off, nevermind *eat* it? But it's true.

The parsnip came to North America on the ships that carried European colonists here and, like the colonists themselves, quickly became naturalized in its new environment. Apparently the feral cousin reverted to producing the nasty chemical armor that had been carefully bred out of its tame and demure relations. So armed, it started taking over fields with a vengeance. Since nothing but the black swallowtail butterfly larvae eat it and nothing but that tough little insect wants to get near it, it's practically invulnerable in the wild.

The domestic relation of the wild parsnip is actually extremely nutritious. With the appearance of pale carrots, they're a nutrient-packed root vegetable with a touch of spice, nuttiness, and sweetness. These vegetables can vary in color from white to cream to pale yellow, with more noticeable sweetness when harvested after the first frost. A half-cup of raw parsnips has about 17 milligrams of vitamin C, about 28% of your daily recommended intake. Boiling parsnips reduces their vitamin C content, but they still provide about 13 milligrams, or 20% of your daily requirements.

But that's the domestic variety. The wild type has no intention of being eaten whatsoever. This former food crop is now a noxious weed that fills roadsides, abandoned fields, railroad embankments, and fencerows across the continent.

If You Love It: May I suggest you do so from a distance?

If You Loathe It: Then grab your shovel and your gloves.

For the Home: Plants can be uprooted or cut down before seeding, but be careful not to expose your skin or eyes to its juices. Cut the younger rosettes below the soil surface with a hoe. For older plants, dig it up with a shovel. If the plant has seed heads, cover them in trash bags and dispose of them. You really don't want to spread this stuff.

For Larger Spaces: The best thing to do is mow the parsnips down with a tractor or riding mower. Be sure to wear protective clothing, and wash thoroughly afterwards. Remember, the phototoxin in the sap reacts with UV light exposure. Do your mowing by June at the latest; you don't want to mow the plants down only to spread the seeds.

If you get sap on your skin:

- Wash your skin thoroughly with soap and water as soon as possible.
- Protect your exposed skin from sunlight for at least 48 hours.
- If you experience a skin reaction, call your healthcare provider.

Rehabilitation and Monitoring: Wild parsnip invades disturbed areas, so control is much more successful when heavily infested areas are re-planted with native tree and plant species that are able to outcompete the seedlings. Reach out to your local extension for ideas on what to plant. Follow-up monitoring and removal of the inevitable new rosettes is absolutely crucial for the next 3 to 5 years, until the seed bank is depleted.

The Chemical Option: Spraying the young rosettes with glyphosate is very effective. Mature plants can't reliably be sprayed, given their size. If you are aware that the site is infested with wild parsnip seeds, a pre-emergent is a good idea where it's practical. In crop situations, go with dicamba or another selective herbicide.

Yellow Woodsorrel

Oxalis stricta

Common Names: Yellow shamrock, sour trefoil, cuckoo bread, Indian sorrel, sheep clover, sheep sorrel, sourgrass, woodsorrel, ladies' sorrel, oxalis, lemon clover, pickle plant, fairy bells.

Family: Oxalidaceae
Genus: *Oxalis*

Looks Like: a dainty nosegay of thin, clover-like leaves, folded like tiny umbrellas on gray days or darkness. The plant grows to about 6 inches high, with tri-compound leaves, heart-shaped leaflets and flowers rising on threadlike stems. The plant flowers from March until the first frost; tiny, 1/2 inch wide flowers have 5 petals and a bright yellow color, making them look like starbursts in their favorite shady corners. Each flower lasts a few days.

The flowers mature into long, skinny seed pods which split dramatically, expelling their tiny seeds in all directions with a little 'pop!' This goes on all through the growing season. They rise from slim, easily broken roots attached to rhizomes deep in the soil, which will immediately get to work sending up leaves to replace the loss. As it grows, its long, multibranched stems eventually flop over and trail out along the ground for as much as 2 feet, extending rootlets from nodes along the stem.

Comes From: Europe.

Likes To Live: in moist, shady, out of the way places with lots of ambient moisture.

217

States Present: Everywhere but the far West.

States Impacted: None in particular.

Legend
- List only
- Law only
- Both List and Law
- Included on other source
- Not Listed

218

Succeeds By: patiently, quietly spreading its rhizomes, and shooting its seeds much less quietly, in all directions. Its bursting seed pods can eject the seeds 10 to 15 feet from itself. A single oxalis plant can produce up to 5,000 seeds per year and have an almost 100% germination rate.

Causes Trouble By: being a bit of a pest and causing plantings or badly tended lawns to look scruffy, but it does no real harm.

Due to a pleasantly bright, sour taste, yellow woodsorrel is sometimes referred to as sour-grass. Its Latin name derives from the Greek word "oxus," which means sour. The sour nature of this weed can be attributed to oxalic acid present in the members of its family, Oxalidaceae.

Speaking of words, there's a good chance that this little beauty is the original shamrock, or seamróg, which is an Irish compound word; the Irish word for clover is seamair, and óg means "young" or "little". Thus, a seamróg is a little clover, and that name fits this shy little plant perfectly.

Legend has it that the great Saint Patrick used the three-leaf clover as a symbol to represent the Holy Trinity when explaining the concept to the pagan Irish. The Irish placed great emphasis on the number 3. For them, it represented life in balance between birth, life and death, and the world in the three elements of earth, sea, and sky. With their regard for the number, it was only natural that Patrick would use a plant with a tri-partate leaf structure to explain the Holy Trinity. It's been the symbol of Ireland ever since; both pagan and christian, both beloved plant and pest.

It's been a classic European spice, adding zip and zing to salads with its little frisson of oxalic acid. Nobody is completely sure when it came to America, but we know it spreads through the horticultural trade; the warm, damp environment under planting benches and around greenhouses is their ideal environment, and many a bedding plant has arrived with woodsorrel seeds hitching a ride in the soil or a little seedling peeping shyly out among its roots.

If You Love It: Then welcome it to some shady corner of your garden, and let it thrive. Fresh yellow woodsorrel leaves and flowers are wonderful as a garnish in a salad, giving it a tangy flavor. Watch how much you eat; high levels of oxalic acid cause kidney stones. But a little makes a wonderful tart treat. Except for the root, the plant can be chewed raw to quench thirst. The green pods can be eaten raw; they have a juicy, crisp texture with a sour flavor similar to rhubarb. The leaves have been used to make a flavored drink that is similar in taste to lemonade, and the whole plant can be brewed as a tea. Due to the oxalic acid and potassium oxalate content, people are warned to limit their intake. But a bit once a week will do you no harm.

It also makes a wonderful groundcover; personally, I have a sneaking soft spot for it and let it grow where I can, watching it fill the difficult bare areas under trees and in terribly shaded spots, with a smile.

If You Loathe It: This fragile perennial isn't hard to pull, but it is very hard to beat.

For the Home

- **Give your grass the chance to beat it:** Growing a thick, healthy lawn is the first line of defense against these mint cousins, since the grass will easily outcompete the weeds for nutrients and growing space when it's healthy. If you've got a spot in the yard where the weed beats the grass, find out the reason why. Sometimes the shade a tree casts or a low spot that holds water makes it hard for the grass. This is when you need a special grass blend.
- **Pull it:** They come out of the ground very easily when young, and can't readily resprout from root fragments.
- **Hoe it:** For areas of young seedlings, hoe them all up by scraping a shuffle hoe over the ground.
- **Mulch it:** A good covering of mulch will prevent the seeds from germinating.
- **Eat it!** Pull the plants up wherever you find them and use them in the kitchen. Just make sure they don't have pesticide residue on them.
- **Accept it:** It really doesn't do any harm; if nothing else will grow where it's growing, why not treat it as a groundcover?

For Larger Spaces

- **Hoe it:** Breaking the little plant out of the ground kills it quickly.
- **Till and Cover it:** Tilling the property and putting in a strong cover crop is a great way of outcompeting the weed.

The Chemical Option: A spraying of 2,4-D is very effective.

Be aware, yellow woodsorrel is doing what nature does best when it colonizes bare ground; protecting the soil. If you don't want weeds on an empty area, either plant something you do want or cover the earth. Nature does not like bare soil. Little yellow woodsorrel is doing all that it can to help the land heal. Don't fight with the land about this; help it.

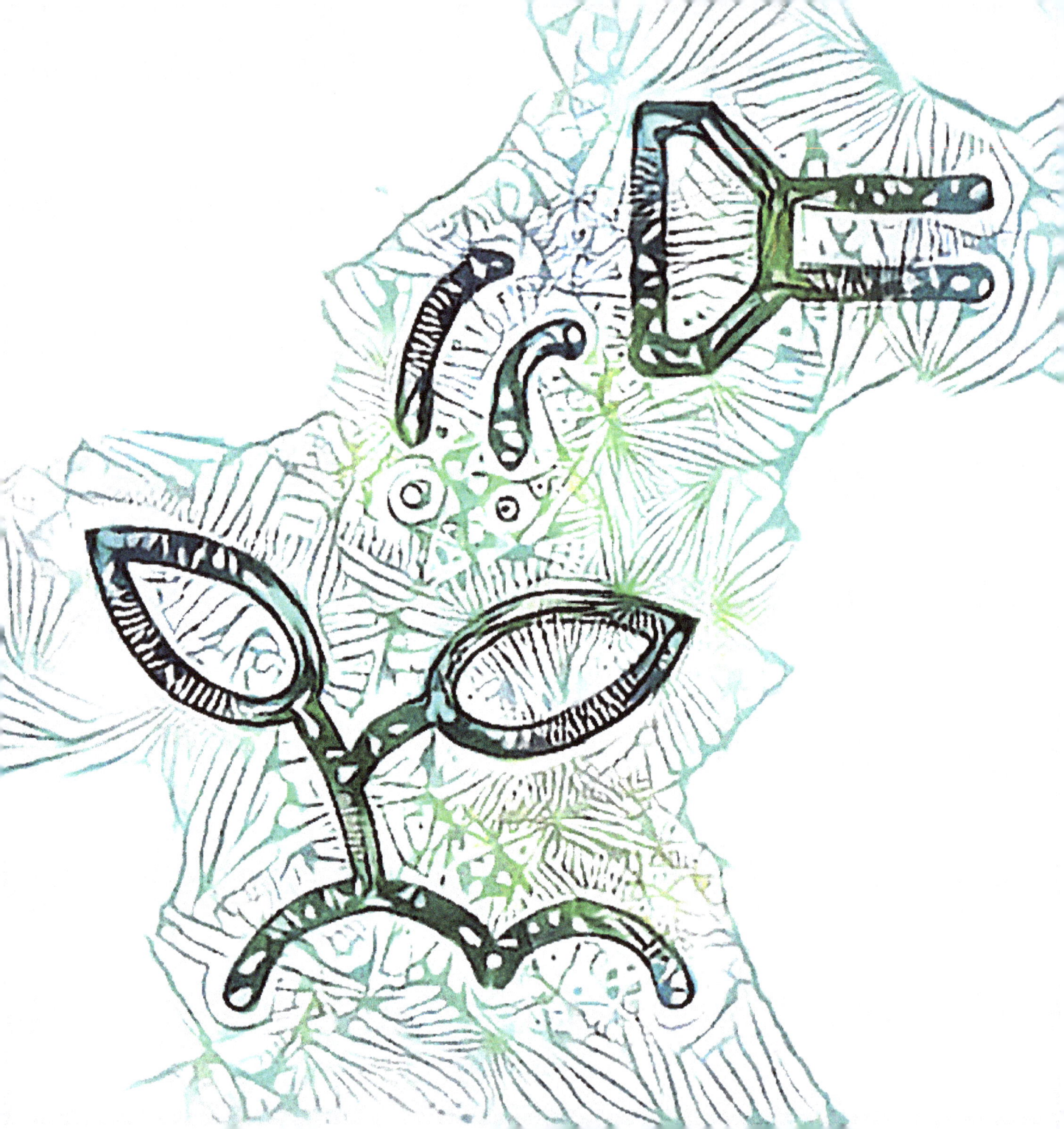

Index Of Plant Terms

Aerial: Referring to the aboveground parts of a plant.

Alternate: A pattern of leaf attachment with leaves appearing one at a time, alternating back and forth down the stem.

Annual: A plant that lives 1 year.

Biennial: A plant that lives 2 years, usually flowering and fruiting in the second year.

Branching Venation: A pattern where leaf veins spread outward like a net or web.

Bulb: An underground storage organ formed by a stem and leaf bases.

C4 Pathway/plant: C4 ("four-carbon") plants initially attach CO_2 to PEP (phosphoenolpyruvate) to form the four-carbon compound OAA (oxaloacetate) using the enzyme PEP carboxylase. This takes place in the loosely packed cells called mesophyll cells. OAA is then pumped to another set of cells, the bundle sheath cells, which surround the leaf vein. By concentrating CO_2 in the bundle sheath cells, C4 plants minimize photorespiration. C4 plants include corn, sugar cane, and many other tropical grasses.

CAM Pathway/Plant: ("crassulacean acid metabolism") plants also initially attach CO_2 to PEP and form OAA. However, instead of fixing carbon during the day and pumping the OAA to other cells, CAM plants fix carbon at night and store the OAA in large vacuoles within the cell. This allows them to have their stomates open in the cool of the evening, avoiding water loss. CAM plants are more common than C4 plants and include cacti and a wide variety of other succulent plants.

Compound Leaf: A single leaf that is broken up into multiple leaflets. 2 common patterns of compound leaf are palmate (hand-like) and pinnate (feather-like).

Cotyledon: The first leaf or leaves produced by a seed.

Crown: A root crown is the part of a root system from which a stem arises. Since roots and stems have very different vascular anatomies, major vascular change takes place here.

Deciduous: A plant that loses its leaves in a specific season.

Dicot: Also called dicotyledon. One of the 2 major divisions of flowering plants. Dicots have 2 cotyledons (seed leaves). Mature leaves have branching leaf veins.

Entire: A leaf-margin pattern where the leaf edges are smooth.

Evergreen: A plant that keeps its leaves in winter.

Forb: A flowering plant other than a grass.

Herbaceous: A broad leafed plant.

Glabrous: Having no hairs or hair-like projections.

Leaf Axil: The area where the leaf joins the stem.

Leaf: A plant structure that is attached to a stem, and which performs photosynthesis and respiration functions.

Leaf Blade: The broad portion of a leaf.

Leaf Shape: The overall shape of a leaf, of which there are many possibilities. Some of the more common leaf shapes include cordate (heart-shaped), round, lance-shaped (lanceolate), palmate, and needle.

Lobed: A pattern of leaf margin where the leaf edge has rounded projections or divisions.

Margins: The edges of a leaf. Common margins include entire, lobed, spiny, toothy, and wavy.

Midrib: The central vein in a leaf.

Monocot: Also called monocotyledons. One of the 2 major divisions of flowering plants. Monocots have a single cotyledon (seed leaf). Mature leaves have parallel leaf veins.

Node: A point along a plant's stem at which one or more leaves or branches can form. Certain plants can also grow roots at nodes.

Opposite: A pattern of leaf attachment. Leaves appearing in pairs, one on each side of the stem.

Palmate: Coming from a central point and spreading out like fingers on a palm. Can refer to leaf shape, leaf venation, or arrangement of leaves or leaflets.

Parallel Venation: A pattern of leaf venation where the veins run alongside one another, without crossing.

Pedicel: The short stalk that a single flower grows on when it is one in a cluster of flowers. The pedicel is attached to the peduncle.

Peduncle: The main stem that supports the pedicels of a group of flowers, or a single flower when there is only one on the stem.

Perennial: A plant that lives more than 2 years.

Petals: The showy, attention-getting parts of a flower.

Petiole: The stalk which connects a leaf to a stem.

Pinnate: Literally meaning "like a feather." It can refer to leaflet arrangement or a venation pattern.

Pistil: The female part of a flower.

Pith: The soft, spongy material in the center of a stem.

Pubescent: Having hairs or hair-like projections.

Raceme: An arrangement of flowers along a stem in which the flowers grow singly on short stalks (pedicels) that are arranged equally along the stem. The flowers open in succession from bottom to top.

Rhizome: An underground, horizontal stem, often confused with a root.

Root: Plants structures that exist for support and water and mineral absorption. Usually underground.

Rosette: A circle of leaves around a central point at, or very close to, ground level.

Sepals: Structures behind the petals. They often serve a protective role.

Sessile: A leaf that directly connects to a stem, without a petiole.

Simple Leaf: Each leaf is a complete leaf unto itself. The leaf blade is not divided.

Spiny: A leaf-margin pattern where the leaf edge has sharp, hard, needlelike projections.

Stamen: The male part of a flower.

Stem: The main trunk of a plant. It provides support and transports resources back and forth across the plant.

Taproot: A large, central root.

Toothy: A leaf-margin pattern where the leaf edge has small jagged projections, similar to teeth or a saw blade.

Venation: The pattern of the veins in a leaf. The 2 main categories are reticulated (branching) and parallel.

Wavy: A leaf margin pattern where the edge of the leaf undulates, as a wave.

Whorl: A pattern of leaf attachment. Leaves appear in groups of 3 or more, around a stem.

Control Measures - Index and Safety Guide

There are a series of steps to handling every pest: identify, learn the life cycles, learn the environment, establish an action threshold, choose the best tactics, implement chosen tactics, then evaluate the outcome. We've discussed a lot of control measures in this book. In this section, let's walk through each of them in detail.

Identify
Properly identifying the problem is key! Your local extension agent is a big help, and there are a number of reliable groups online.

Learn the life cycles
Knowing when and how to go after a pest depends on knowing when and how it's vulnerable.

Learn the environment
To understand the scale of what you're dealing with, watch the environment and discern the scope of the damage involved. Sometimes you'll realize that the pest-plant is simply that; a pest, not a danger.

Establish an action threshold
This is another way of saying, "How much am I willing to lose?" This will depend on your needs and your circumstances. Maybe at your house, a few weeds under the rose bushes are no problem, but a poisonous plant must be eradicated when it grows near areas where children play.

Choose the best tactics
Choosing the best combination of tactics is the key to developing your approach.

Implement chosen tactics

Use all the proper materials and carefully observe all the relevant safety measures as you get your infestation under control. If you are using herbicides, be careful to rotate the products. To help delay the development of herbicide resistance, do not use herbicides with the same mode of action year after year. That will simply act as a selective pressure leading to resistance.

Evaluate the outcome

Evaluating the results is absolutely essential. If your plan to control your pest didn't work in some way, go back through the steps and figure out where it went wrong, or where you misidentified an element of the problem. Don't assume that the project didn't work if your pest didn't die after one attempt; in some cases, as with tree of heaven or bindweed, the problem with 'I sprayed it and it just came back' wasn't the spray, but the expectation that the plant would die on the first application!

Integrated Pest Management

Integrated pest management is the best approach to handling weeds, and other garden pests as well. Picture this system as a pyramid, with the most important element as the base.

Pyramid Base - Cultural Control

The solid base of the pyramid should always be cultural control. Cultural control is the practice of reducing weed and pest populations and their survival by choosing the right plants and gardening practices for your environment. Examples of cultural control include a good mowing schedule, water-wise irrigation and watering, and care for both your soil and plants.

Maintain good watering practices

Water according to your conditions, not your schedule. In areas with mildew issues, water in the mornings, so that moisture can evaporate during the daylight hours. This also avoids leaf burn caused by drops of water acting like tiny magnifying glasses in the sun. However, in high-heat situations, night watering maximizes the water absorption. Many managers split the difference and water around 3 or 4 in the morning.

Make sure you regularly walk your property and check the effectiveness of your watering. Look for dry spots and soggy spots; these conditions are both unhealthy for many cultivated plants. Improve drainage or cut down your watering to remove soggy areas; increase watering for areas that are dry. And please, please check your system. You really don't want to be the person running the irrigation when it's raining or watering the sidewalk instead of the lawn. It wastes water, it wastes money, and it isn't going to make you any friends in the neighborhood.

Plant wisely
Choose the right plants for your site and care for them in accordance with their needs; healthy plants can outcompete most weeds.

DO NOT TAKE SHORTCUTS
Although a short-term solution to a problem may have satisfying, immediate results, these shortcuts lead to long-term headaches. Don't plant large plants in small spaces to make your property look finished faster, for example. And don't plant a species you like in a site that isn't suited for it. That's a road to heartbreak. Good cultural practices involve planting species naturally resistant to diseases, pests, and infections, and avoiding those that are susceptible to trouble.

Pyramid Second Level - Mechanical control

Mechanical control is the second level of the pyramid. Mechanical control is the use of anything that makes the weed's environment unwelcoming, or kills it.

Pulling and Tilling
Removing the plants from the ground is often the quickest, easiest and most effective method to weed suppression in small areas. Here's a few tips.

Don't put off your weeding. If you let weeds get the upper hand, you'll have a tough time getting them out. When weeds are small, their roots are too; that's when you want to pull them. Commit to doing a quick walk-through of your garden every other day; it will take only a few minutes to pull up any young weeds that show up.

Grab plants by the base. Remember to wear your gloves! Gardeners who weed manually will sometimes reach down, grab the nearest leaves at hand and tug. Unfortunately, that often causes the weed they're tugging on to snap in two, leaving the bottom half and the roots still in the ground. Take your time and grab each weed individually at its base, then pull slowly and steadily to ease the roots from the soil. Tools are also a great idea.

Use the right tools, at the right time. Hand tools can make your life a lot easier and more comfortable. Choose well-made tools with a solid handle that feels comfortable in your grip, and a head or blade made of forged steel.

● **Kneeling tools:** I personally prefer a soil knife for daily work. It has the appearance of a cross between a trowel and a short sword, and it has a notch in the tip to get under and lever up stubborn weeds. To remove weeds between beneficial plants, try an angled hand hoe.

● **Standing tools:** For removing large weeds or a large number of weeds at once, the long-handled hoe is the best approach. The point of this tool isn't to chop up the soil, though some people try to use it that way. Instead, set the blade at the base of the plant and then swing, as if the weed was a golf ball. Alternately, use a circle or shuffle hoe to shuffle the roots out of the ground; this causes less disturbance to the soil. Really, anything that can sever roots beneath the soil surface will work wonders. It also does less damage to your back than kneeling all day with a soil knife in hand.

For the really pernicious weeds, a good shovel is your friend. Dig as deep and get as much of the roots as you can. Remember never to overextend your body; you generally want your spine to stay in line with your knees, otherwise you can hurt your back, one knee, or both. If you bend your knees or your back into an acute angle, that's a problem.

- **Tilling:** This can be done by hand, or with machinery in large areas. The goal is to break up the roots of plants where that's an effective way of killing them. A four-pronged cultivator works well for hand-work. For large spaces, there are a number of tractor-pulled options. **Remember not to till plants that can spread by root or stem fragments!**

- **Mulch:** At a home, mulching and hand-pulling is usually a very good solution. Weeds need leaves to synthesize food, so if you take away their ability to produce the food necessary to feed the roots, you will slowly and surely kill the troublemakers. Seeds need light to germinate, so mulch over bare dirt is a very effective way of preventing them from getting a headstart. Bark, compost, or straw laid at least 2 inches thick can effectively control many weed seeds. A large, coarse bark will require a 3 to 4 inch layer to be effective; however, larger, coarser mulches last longer than finely shredded ones. Thick mulch eventually can accumulate soil, decaying organic matter, and weed seeds that can germinate. All organic mulch needs periodic replacement, but it also feeds your soil over time, so that's no problem.

- **Weed Barriers:** Synthetic polypropylene weed barriers (fabrics or geotextiles), which are available at nurseries, also block sunlight and starve weed seedlings. The fabrics are porous to allow water to drain through them. A synthetic barrier with bark or rock on top makes the area more aesthetically pleasing. Organic mulches such as bark and straw don't need to be as thick if you also are using the fabric. Since mulches and weed barriers reduce evaporation from the soil surface, adjust the irrigation cycle to prevent overwatering.

However, you do need to be aware that weed barriers will degrade over time, and replacing them is a monumental task. Poorly cared for weed barrier poking out through scanty mulch is one of the least attractive things you'll ever see, so consider this when you're choosing whether to use it. Personally, I place it under rock in areas that are intended to be without plantings in my designs.

- **Fire:** Another very effective mechanical approach is the fire-wand. When flame-weeding, the most effective method is to catch weeds early. When they are small, flaming is almost completely effective at killing non-native weeds; plants over 4 inches are more difficult to kill without multiple flamings.

  A tip: You do not want to burn weeds to ash! Just a split second of heat should kill unwanted weeds and grasses completely; you don't need to burn them to a crisp. For best results, increase exposure to the heat if weeds are wet from dew. Water on the leaves acts as insulation and decreases cell damage unless you run the flame over them a bit longer. Note: it is ok to flame when it is wet out. In fact, we recommend it. Moisture will lessen the threat of ignition of dry debris.

  Please be aware of the fire danger in your area. Don't use flame on a windy day, or in high heat. For larger prescribed burns, please consult with your local fire authority and agricultural extension agent. You really don't want to be the cause of the next wildfire in the news. Schedule your burn carefully too: you don't want to provide an open, newly fertilized space for weeds that are usually suppressed in an area to take off!

- **Solarizing.** Covering the soil with sheets of clear plastic for 4 to 6 weeks during the summer can effectively reduce the number of seeds in areas where summer daytime temperatures are very hot. In areas where summer temperatures are lower than 90°F, soil solarization can partially control weed seeds.

- **Biological controls.** The third step of the pyramid is the introduction of a biological control agent. Biological control involves the use of natural predators to control pests. Biological control agents can be useful in the right situation. Successful examples include weevils released to control puncturevine. Since biological controls are extremely species-specific and this is an evolving field, you'll need to select the proper predator for the site and the pest. Never release a biological control organism in temperatures they can't survive; remember, these are living things. Act accordingly.

- **Chemical Control:** This includes the use of any sprays, powders, granules, and any other chemical applied to kill, repel, or suppress weeds. Chemicals are the last option in your list of possibilities. Spraying chemicals in the garden should always be done with safety protocols in mind. The most important element of using any chemical is reading the label. Be aware, **the label on any type of pesticide is a legal document**. You are legally bound to abide by its safety protocols, and misusing the product makes you culpable.

Safety Protocols and Best Practices

When applying anything in the garden, always act with safety in mind. Let's be frank for a moment: cutting corners with safety is stupid. It's bad for you and it's bad for your garden. Don't do it.

Safety Clothing

Basic levels of protection are important while applying pesticides. Your goal is to avoid contact with any pesticide on any portion of your skin. Every product you use requires different safety protocols and equipment. Each application is different, but there are a few basic practices and pieces of gear that are required for almost all applications.
- **Chemical gauntlets:** Thick nitrile gloves that reach at least as high as the forearms.
- **Close-toed boots:** Preferably a good pair of boots you don't use on a lot of other things.
- **Covering clothing:** I use jeans and a button-up shirt myself. Some people have specific spray jackets.

Best Practices

Spray during dry, calm weather under 80 degrees. Cover leaf surfaces thoroughly with spray droplets, but do not spray to the point that liquid is dripping off the leaves. It may take plants treated up to a week or more, depending on moisture levels, before symptoms appear. Do not cut or dig the plant until the top growth has died back. If the leaves remain green a month after the initial treatment, check that the product has not expired and spray them again.

Be aware of drift. "Drift" is the term used to describe what happens when your application hits unintended areas. This can happen through a casual movement with the spray wand, a sudden gust of wind, or it can occur through volatilization: on hotter days, many chemicals can become gasses that drift over unintended areas. Spraying a non-selective herbicide carelessly is a recipe for disaster, especially when the wind blows the spray onto unintended targets. Spraying herbicides on a hot day is asking for them to volatilize and harm unintended plants. Don't spray when it's over 80 degrees outside!

Make sure you don't paint yourself into a corner, either. Imagine spraying a non-selective herbicide on a patch of weeds, walking through it, and then tracking the herbicide through your lawn.

If you are working near water, choose only chemical formulations that are aquatic-safe. Terrible things happen when harmful chemicals end up in streams or ponds. Don't be the person who caused such a disaster. Read the labels, and use an aquatic-safe formulation.

Be mindful when mixing chemicals. This is when most spills and accidental contacts occur. Follow the directions on the label and avoid using water directly from a hose. Never, ever mix the chemicals near a drain, and for the love of ecology, don't dump the rinse water in the street!

Basic Guide to Chemical Formulations

In general, most homeowners are going to interact with either liquid or solid formulations.

Liquid Formulations

Most pesticide applications involve spraying a liquid chemical to achieve the desired effect.

Some combinations are available in a pre-mixed container; in this case, read the label and apply as instructed. At times, you may find concentrates of particular products; you add a small amount of concentrate to a portion of water and mix before application. Concentrates are more effective if you plan on repeated applications.

Liquid formulations usually take effect immediately, but require regular reapplications. These liquids can be hazardous if they come into contact with you in any way. Good choices include 2,4-D (Trimec, Triplet, End Run), dicamba (Banvel, or Vanquish), glufosinate (Finale, Liberty, or Rely), glyphosate (Roundup), and paraquat (Gramoxone). Be aware that glyphosate is a known carcinogen!

Vinegar-based sprays are another organic/all-natural solution to eliminating weeds and undesirable plants in the garden. Any of these sprays, homemade or purchased, will work to kill surface leaves of most plants but do not kill the roots. Frequent application is necessary when using a vinegar-based spray.

Solid formulations

Granules, powders, and similar products are commonly used. They are useful as pre-emergents to control the growth of weeds, as well as provide slow-release nutrients.
Solids are beneficial, because they provide longer lasting impact. Most granules will slowly break down over time and through exposure to water, providing a steady, measured release. By reading the directions on the label, you'll learn the expected duration of these products and will be provided with a measured calendar of application.
In general, these formulations are the most dangerous to your eyes, mouth, and nose. Wear safety goggles, gloves, and a mask. Before applying any pesticide, read the label and become familiar with the necessary actions to take if you've been exposed.

A particularly useful solid formulation is something in the family of pre-emergents, the best tool in the fight against prolifically seeding weed seeds in mature plantings. To properly use pre-emergents, time the application so it occurs before the soil temperature exceeds 55° to 60°F at a depth of 1 inch.

Pre-emergent herbicides for turf and ornamentals include benefin (Balan), pendimethalin (Pendulum), isoxaben (Gallery), oryzalin (Surflan), trifluralin (Treflan, Preen), and dithiopyr (Dimension). The most effective pre-emergent herbicides are atrazine (Aatrex), bromacil (Hyvar), chlorsulfuron (Telar), hexazinone (Velpar), imazapyr (Arsenal), napropamide (Devrinol), simazine (Princep), and sulfometuron (Oust). Of these, only pendimethalin, trifluralin, dithiopyr, and oryzalin are available for use by home gardeners. Combination products such as oryzalin plus benefin are available to both home gardeners and landscape professionals.

Preventing problems from arising in the first place is the real secret to weed control. When it comes to using fire or applying chemicals, it's vital to practice diligent safety practices.

Bibliography

Bindweed

- Austin, Daniel F. "Bindweed (*Convolvulus arvensis*, Convolvulaceae) in North America, from medicine to menace." Journal of the Torrey Botanical Society (2000): 172-177.
- Invasive Plant Atlas of the United States [Online.] Available: https://www.invasiveplantatlas.org/subject.html?sub=4338
- Boydston, R. A. and M. M. Williams. 2003. Combined effects of *Aceria malherbae* and herbicides on field bindweed (*Convolvulus arvensis*) growth. Weed Sci. 52(2):297-301.
- Guntli, Daniel, et al. "Biological control of hedge bindweed (*Calystegia sepium*) with *Stagonospora convolvuli* strain LA39 in combination with competition from red clover (*Trifolium pratense*)." Biological Control 15.3 (1999): 252-258.
- Lyons, K. E. 1998. Element stewardship abstract for *Convolvulus arvensis* L. Field bindweed. [Online.] Available: http://tncweeds.ucdavis.edu/esadocs/documnts/convarv.pdf. Prepared for The Nature Conservancy, Arlington, VA 21 pp.
- Mitich, L. W. 1991. Intriguing world of weeds: field bindweed. Weed Technol. 5:913-915.
- Mitich, Larry W. "Field bindweed." Weed Technology 5.4 (1991): 913-915.
- Pfiriter, Hanspeter A., et al. "Towards the management of field bindweed (*Convolvulus arvensis*) and hedge bindweed (*Calystegia sepium*) with fungal pathogens and cover crops." Integrated Pest Management Reviews
- Roth, Sally. Weeds, Friend Or Foe?. Reader's Digest, 2002.
- Zollinger, R. K. and R. G. Lym. 2000. Identification and control of field bindweed. North Dakota State Univ. Ext. Ser. Circ. W-802. Fargo, ND.
- EDDMapS. 2021. Early Detection & Distribution Mapping System. The University of Georgia - Center for Invasive Species and Ecosystem Health. Available online at http://www.eddmaps.org

Black Medic

- Carlisle, Liz. Lentil Underground: Renegade farmers and the future of food in America. Avery, 2016.
- Deane, G. Eat The Weeds, http://www.eattheweeds.com/black-medic/
- Duke, J. A. and E. S. Ayensu 1984.Medicinal Plants of China. Algonac, Michigan, Reference Publications, Inc.
- EDDMapS. 2021. Early Detection & Distribution Mapping System. The University of Georgia - Center for Invasive Species and Ecosystem Health. Available online at http://www.eddmaps.org
- Great Plains Flora Association. 1986. Flora of the Great Plains. Lawrence, KS: Univ. Press of Kansas. 1392 pp.
- Mitich. L. W. 1983. Intriguing world of weeds - bugseed, Indian hemp, common milkweed, Black medic.

Weeds Today. 14(3):8-10.
- Pavone, L. V. and R. J. Reader. 1985. Reproductive schedule of *Medicago lupulina* (Leguminosae) in a patchy environment. Can. J. Bot. 63:2044-2048.
- Smith, R. C. and R. Zollinger. 1993. Weed control in North Dakota turfgrasses. North Dakota State Univ. Ext. Serv. Circ. H-1009. Fargo, ND.
- Turkington, R. and P. B. Cavers. 1979. The biology of Canadian weeds. 33. *Medicago lupulina* L. Can. J. Plant Sci. 59:99-110.
- Whitson, T. D., editor. 2000. Weeds of the West 9th Ed. Western Society of Weed Science, Newark, CA

Bradford Pear

- Adrian Higgins, 'Scientists thought they had created the perfect tree. But it became a nightmare', Washington Post September 17, 2018. https://www.washingtonpost.com/lifestyle/magazine/how-we-turned-the-bradford-pear-into-a-monster/2018/09/14/f29c8f68-91b6-11e8-b769-e3fff17f0689_story.html
- Alabama Cooperative Extension System, https://www.aces.edu/
- Alabama Department of Conservation and Natural Resources
- Ann Koenig, STOP THE SPREAD! Missouri Department of Conservation Feb 17, 2011
- Culley, Theresa M., and Nicole A. Hardiman. "The beginning of a new invasive plant: a history of the ornamental Callery pear in the United States." BioScience 57.11 (2007): 956-964.
- EDDMapS. 2021. Early Detection & Distribution Mapping System. The University of Georgia - Center for Invasive Species and Ecosystem Health. Available online at http://www.eddmaps.org.
- Indiana Department of Natural Resources
- https://www.invasive.org
- Missouri Botanical Garden, https://www.missouribotanicalgarden.org/
- Missouri Department of Conservation
- Ohio Invasive Plants Council

Burdock

- Culpeper, Nicholas. The English Physitian : OR AN Astrologo-Physical Discourse of the Vulgar Herbs of This Nation. London. England: Printed by Peter Cole, at the sign of the Printing-Press in Cornhill, near the Royal Exchange., 1652.
- Dioscorides. De Materia Medica. - Five Books in One Volume: A New English Translation. Translated by Osbaldeston, T. Johannesburg: IBIDIS Press, 60AD.
- Dyer, Mary H. , Burdock Management: Tips For Controlling Common Burdock Weeds https://www.gardeningknowhow.com/edible/herbs/burdock/controlling-burdock-weeds.htm
- EDDMapS. 2021. Early Detection & Distribution Mapping System. The University of Georgia - Center for Invasive Species and Ecosystem Health. Available online at http://www.eddmaps.org
- Hatfield, Gabrielle. Encyclopedia of Folk Medicine: Old World and New World Traditions. Santa Barbara, CA: ABC-CLIO, 2004.
- Invasive Plant Atlas of the United States, https://www.invasiveplantatlas.org/subject.html?sub=5140

- Invasive Plants Fact Sheet - Common Burdock https://www.fs.usda.gov/naspf/resources/invasive-plants-fact-sheet-common-burdock
- Lloyd, John Uri. Origin and History of All the Pharmacopeial Vegetable Drugs, Chemicals and Preparations with Bibliography. Vol. 1. Cincinnati, OH: Caxton Press, 1921.
- Morrell, Jennie MH. "Some Maine Plants and Their Uses 'Wise and Otherwise.'" Rhodora 3, no. 29 (1901)
- Ody, Penelope. The Complete Guide to Medicinal Herbal: Herbal Remedies for Common Ailments. Dorling Kindersley, 1993.
- Wooster. The American Practice of Medicine. Vol. III. 3 vols. New York, NY: Reformed Medical School Society, 1833.

Canada Thistle

- Alberta Invasive Plants Council (Canada).
- Botkin, Benjamin Albert, ed. A treasury of American folklore. New York: Crown, 1944
- https://crops.extension.iastate.edu/encyclopedia/brief-history-canada-thistle-iowa
- Eckberg, J., et al. "Native Thistles: A Conservation Practitioner's Guide." Plant Ecology, Seed Production Methods, and Habitat Restoration Opportunities (2017)
- EDDMapS. 2021. Early Detection & Distribution Mapping System. The University of Georgia - Center for Invasive Species and Ecosystem Health. Available online at http://www.eddmaps.org
- Fact Sheet: Canada Thistle (Jan 2014; PDF | 481 KB)
- Hayden, A. 1934. Distribution and reproduction of Canada thistle in Iowa. American Journal of Botany 21: 355-373.
- Mack, Richard N. "Plant Naturalizations and Invasions in the Eastern United States: 1634-1860." Annals of the Missouri Botanical Garden, vol. 90, no. 1, 2003, pp. 77–90. JSTOR, www.jstor.org/stable/3298528. Accessed 15 Feb. 2021.
- Nuzzo, V. 1997. Element Stewardship Abstract for Cirsium arvense, Arlington, Va.: Nature Conservancy.
- Polley, Jane, ed. American folklore and legend. Reader's Digest, 1978
- Roth, Sally. Weeds, Friend Or Foe?. Reader's Digest, 2002.
- Royer, F. and R. Dickinson. 1999. Weeds of the Northern U.S. and Canada. The University of Alberta press. 434 pp.
- Stachion, W.J. and R.L. Zimdahl. 1980. Allelopathic activity of Canada thistle (Cirsium arvense) in Colorado, Weed Science 28:83-86.
- Trumble, J. T., & Kok, L. T. 1982. Integrated pest management techniques in thistle suppression in pastures of North America. Weed Research, 22(6):345-359.

Couch Grass

- The American physitian, or, A treatise of the roots, plants, trees, shrubs, fruit, herbs, &c. growing in the English plantations in America : describing the place, time, names, kindes, temperature, vertues and uses of them, either for diet, physick, &c. ; William Hughes. 1672.

- Batcher, Michael S., "Element Stewardship Abstract for Elytrigia repens var. repens (L.) Desv. ex B. D. Jackson," prepared for The Nature Conservancy, 2002.
- Cholewa, A. F., "Elytrigia," J. F. Bell Museum of Natural History, 2002.
- Claus, J. S., & Behrens, R. 1976. Glyphosate translocation and quackgrass rhizome bud kill. Weed Science, 149-152.
- "Controlling Quackgrass in Gardens," Yard and Garden Brief, University of Minnesota Extension Service, 1999.
- Curran, Williams S., and Dwight D. Lingenfelter, "Quackgrass Management: An Integrated Approach," Penn State College of Agricultural Sciences, 2002.
- EDDMapS. 2021. Early Detection & Distribution Mapping System. The University of Georgia - Center for Invasive Species and Ecosystem Health. Available online at http://www.eddmaps.org
- Ivany, J. A. 2002. Control of Quackgrass (*Elytrigia repens*) and Broadleaf Weeds and Response of Potato (*Solanum tuberosum*) Cultivars to Rimsulfuron 1. Weed technology, 16(2):261-266.
- MSU Crop Advisory Team Alerts, https://www.canr.msu.edu/outreach/
- OARDC-Ohio Perennial and Biennial Weed Guide; https://www.oardc.ohio-state.edu/weedguide
- "Quackgrass and Its Control," Iowa State University, University Extension, April 1992.
- "Quitch (Couch-grass) Root," Flora, http://www.florahealth.com/flora/home/Canada/HealthInformation/Encyclope...
- "Weeds, Friends or Foe?: Unusual Controls," Golden Harvest Organics, http://www.ghorganics.com/page26.html

Crabgrass

- Abaye, Azenegashe Ozzie, et al. "The nutritive value of common pasture weeds and their relation to livestock nutrient requirements." (2009).
- CABI, 2021. Digitaria ciliaris [original text by Lavin, Matt]. In: Invasive Species Compendium. Wallingford, UK: CAB International. www.cabi.org/isc.
- Contreras, Patricia. "NEW JERSEY OPINION; IF YOU CAN'T LICK 'EM, EAT 'EM." May 3, 1981, Section 11, Page 36 of the New York Times
- Corriher-Olson, Vanessa, "Weed of the Week: Crabgrass", Texas A&M AgriLife Extension July 27 2020. https://ag.umass.edu/turf/fact-sheets/biology-management-of-crabgrass
- Deane, Green, "Crabgrass Was King", Eat The Weeds And Other Things Too, March 20 2012, http://www.eattheweeds.com/crabgrass-digitaria-sanguinalis
- EDDMapS. 2021. Early Detection & Distribution Mapping System. The University of Georgia - Center for Invasive Species and Ecosystem Health. Available online at http://www.eddmaps.org
- Gardner, Dave. Agronomy 517: Weed Biology and Ecology: Crabgrass. Spring Semester 1996. http://agron-www.agron.iastate.edu/~weeds/WeedBiolLibrary/u4crab1.html
- Mitich, Larry W. "Crabgrass." Weed Technology 2.1 (1988): 114-115.
- Peters, R. A., and Stuart Dunn. "Life history studies as related to weed control in the Northeast; 6: large and small crabgrass." (1971).
- Temme, D. G., R. G. Harvey, R. S. Fawcett, and A. W. Young. 1979. Effects of annual weed control on alfalfa forage quality. Agronomy Journal 71 (1979): 51-54.
- Weed Science Society Of America, https://wssa.net/wp-content/themes/WSSA/WorldOfWeeds/crabgrass.html

Creeping Bellflower

- A Performance Appraisal of Hardy Bellflowers – Plant Evaluation Notes of the Chicago Botanic Garden at www.chicagobotanic.org/downloads/planteval_notes/no31_bellflower.pdf
- Coburn, Carl W., Albert T. Adjesiwor, and Andrew R. Kniss. "Creeping Bellflower Response to Glyphosate and Synthetic Auxin Herbicides." HortTechnology 28.1 (2018): 6-9.
- Creeping Bellflower – on the Wildflowers of Illinois website at www.illinoiswildflowers.info/weeds/plants/cr_bellflower.htm
- Czarapata, Elizabeth J. Invasive plants of the upper Midwest: an illustrated guide to their identification and control. Univ of Wisconsin Press, 2005.
- EDDMapS. 2021. Early Detection & Distribution Mapping System. The University of Georgia - Center for Invasive Species and Ecosystem Health. Available online at http://www.eddmaps.org/; last accessed January 20, 2021.
- Gilbert, Oliver. The ecology of urban habitats. Springer Science & Business Media, 2012.
- Laska, Vera. The Czechs in America, 1633-1977. Dobbs Ferry, New York: Oceana Publications, 1978.
- Molinari, Christine. Czech americans, Multicultural America: Countries and their Cultures, Available online at https://www.everyculture.com/multi/Bu-Dr/Czech-Americans.html#ixzz6k8Qg3t2d
- Panke, Brendon, Ryan deRegnier, and Mark J. Renz. Creeping bellflower (Campanula rapunculoides). University of Wisconsin—Extension, Cooperative Extension, 2012.
- Plants For A Future: A resource and information centre for edible and otherwise useful plants, https://pfaf.org/user/Default.aspx
- Roquet, Cristina, et al. "Reconstructing the history of Campanulaceae with a Bayesian approach to molecular dating and dispersal–vicariance analyses." Molecular Phylogenetics and Evolution 52.3 (2009): 575-587.
- Traill, Catherine Parr Strickland, 1802-1899, and Fitzgibbon Agnes. North American Wild Flowers. Original printing 1870. Reprint Good Press, 2019
- Writers' Program of the Work Projects Administration in the State of Minnesota. The Bohemian Flats. St. Paul, Minnesota: Minnesota Historical Society Press, 1986.

Dandelion

- Baker Creek Heirloom Seeds, https://www.rareseeds.com/
- Baumflek, Michelle J., Marla R. Emery, and Clare Ginger. "Culturally and economically important nontimber forest products of northern Maine." (2010).
- EDDMapS. 2021. Early Detection & Distribution Mapping System. The University of Georgia - Center for Invasive Species and Ecosystem Health. Available online at http://www.eddmaps.org/; last accessed January 20, 2021.
- Frick, B., and A. G. Thomas. "Weed surveys in different tillage systems in southwestern Ontario field crops." Canadian Journal of Plant Science 72.4 (1992): 1337-1347.
- Gail, Peter A. The dandelion celebration: a guide to unexpected cuisine. Goosefoot Acres Press, 1994.

- Grauso, Laura, et al. "Common dandelion: a review of its botanical, phytochemical and pharmacological profiles." Phytochemistry Reviews 18.4 (2019): 1115-1132.
- Hourdajian, D. "Introduced species summary project dandelion (Taraxacum officinale)." Invasion Biology Introduced Species Summary Project–Columbia University (2006).
- List, Uses, and Host Plant. "Taraxacum officinale complex (dandelion)." Centre for Agriculture and Bioscience, International Invasive Species Compendium (2020).
- Mitich, Larry W. "Common Dandelion–the Lion's Tooth." Weed technology 3.3 (1989): 537-539.
- Ody, Penelope. The complete medicinal herbal: a practical guide to the healing properties of herbs. Simon and Schuster, 2017.
- Reader's Digest Association. Magic and medicine of plants. Readers Digest, 1986.
- Sanchez, Anita. The Teeth of the Lion: The Story of the Beloved and Despised Dandelion. McDonald and Woodward Publishing Company, 2006.
- Wetli, Patty. The Dandelion's Fall From Grace Has Been a Doozy. Can This Weed Become a Flower Again? WTTW News May 15, 2020. https://news.wttw.com/2020/05/14/dandelion-weed-flower-history
- Wirngo, Fonyuy E., Max N. Lambert, and Per B. Jeppesen. "The physiological effects of dandelion (Taraxacum officinale) in type 2 diabetes." The review of diabetic studies: RDS 13.2-3 (2016): 113.

Dock

- Bezusko, Lyudmila G., Timur V. Bezusko, and Sergei L. Mosyakin. "A Reconstruction of the Flora and Vegetation in the Central Area of Early Medieval Kyiv, Ukraine, Based on the Results of Palynological Investigations."
- Bish, Mandy. Weed of the Month: Curly Dock (Rumex crispus). Missouri Extension 2015. https://ipm.missouri.edu/ipcm/2015/4/Weed-of-the-Month-Curly-Dock/
- Blair, Katrina. The wild wisdom of weeds: 13 essential plants for human survival. Chelsea Green Publishing, 2014.
- Dioscorides, Translated by Tess Anne Osbaldeston, De Materia Medica. - Five Books in One Volume: A New English Translation, IBIDIS Press (60AD) p 263
- Gabrielle Hatfield, Memory, Wisdom and Healing: The History of Domestic Plant Medicine (Kindle Locations 1974-1976), The History Press (2012), Kindle Edition
- Green, Aliza. Field Guide to Herbs & Spices: How to Identify, Select, and Use Virtually Every Seasoning at the Market. Quirk books, 2006.
- Lakoba, Vasiliy T., et al. "An Analysis of US State Regulated Weed Lists: A Discordance between Biology and Policy." BioScience 70.9 (2020): 804-813.
- Lavin, Matt. "Invasive species compendium." (2016).
- Nicholas Culpeper, Culpeper's Complete Herbal and English Physician [1981 Reprint], J. Gleave and Son, Deansgate (1826)Pfeiffer, Ehrenfried E. Weeds and What they Tell us. Floris Books, 2016.
- Stephen Pollington, Leechcraft: Early English Charms, Plant Lore, and Healing Anglo-Saxon Books (2008)

Foxtail Grass

- Cathcart, R. Jason, and Clarence J. Swanton. "Nitrogen management will influence threshold values of green foxtail (Setaria viridis) in corn." Weed Science 51.6 (2003): 975-986.
- EDDMapS. 2021. Early Detection & Distribution Mapping System. The University of Georgia - Center for Invasive Species and Ecosystem Health. Available online at http://www.eddmaps.org/; last accessed February 2, 2021
- https://www.seattledogspot.com/foxtail-grass/
- Georgia, Ada Eljiva. A Manual of Weeds: With Descriptions of All the Most Pernicious and Troublesome Plants in the United States and Canada, Their Habits of Growth and Distribution, with Methods of Control. Macmillan, 1914.
- Groh H, Frankton C (1949) Canadian weed survey. 7th report. Canada Department of Agriculture, Ottawa, p 144
- Manson JM (1932) Weed survey of the Prairie Provinces. Dominion of Canada, p 34
- Schröder, Stephan, et al. "Genetic diversity and origin of North American green foxtail [Setaria viridis (L.) Beauv.] accessions." Genetic Resources and Crop Evolution 64.2 (2017): 367-378.
- Schultz, Ryan M., and Allison Zwingenberger. "Radiographic, computed tomographic, and ultrasonographic findings with migrating intrathoracic grass awns in dogs and cats." Veterinary Radiology & Ultrasound 49.3 (2008): 249-255.
- Turbelin, Anna J., Bruce D. Malamud, and Robert A. Francis. "Mapping the global state of invasive alien species: patterns of invasion and policy responses." Global Ecology and Biogeography 26.1 (2017): 78-92.

Giant Hogweed

- EDDMapS. 2021. Early Detection & Distribution Mapping System. The University of Georgia - Center for Invasive Species and Ecosystem Health. Available online at http://www.eddmaps.org/; last accessed February 2, 2021
- Holloway, Cas, and Commissioner Paul Rush. "New York City Department of Environmental Protection." (2010).
- Lavin, Matt. "Invasive species compendium." (2016).
- Mihir Zaveri and Christine Hauser. Giant Hogweed: A Plant That Can Burn and Blind You. But Don't Panic. New York Times July 2, 2018
- Nielsen C, Vanaga I, Treikale O, Priekule I, 2007. Mechanical and chemical control of Heracleum mantegazzianum and H. sosnowskyi. In: Ecology and management of giant hogweed (Heracleum mantegazzianum) [ed. by Pys?ek P, Cock MJW, Nentwig W, Ravn HP] Wallingford, UK: CABI, 226-239
- Nielsen, Charlotte, et al. "The Giant Hogweed Best Practice Manual. Guidelines for the management and control of an invasive weed in Europe." Forest and Landscape Denmark, Hoersholm 44 (2005).
- Pyšek, Petr, et al. Ecology and management of giant hogweed (Heracleum mantegazzianum). CABI, 2007.
- Sampson, Clare. "Cost and impact of current control methods used against Heracleum mantegazzianum (giant hogweed) and the case for instigating a biological control programme." Ecology and management of invasive riverside plants. (1994): 55-65.
- Turbelin, Anna J., Bruce D. Malamud, and Robert A. Francis. "Mapping the global state of invasive alien species: patterns of invasion and policy responses." Global Ecology and Biogeography 26.1 (2017): 78-92.

Japanese Knotweed

- Adams, William M. "Sites of Special Scientific Interest and habitat protection: implications of the Wildlife and Countryside Act 1981." Area (1984): 273-280.
- Bailey,J.P. and A. P. Connolly. 2000. Prize-winners to pariahs—A history of Japanese knotweed s.l. (Polygonaceae) in the British Isles. Watsonia 23: 93–110.
- Balogh, Lajos. "Japanese, giant and bohemian knotweed." The most important invasive plants in Hungary. Vacratot: Institute of Ecology and Botany, Hungarian Academy of Sciences (2008): 13-33.
- Barney, Jacob N. "North American history of two invasive plant species: phytogeographic distribution, dispersal vectors, and multiple introductions." Biological Invasions 8.4 (2006): 703-717.
- Clements, David R., Todd Larsen, and Jennifer Grenz. "Knotweed management strategies in North America with the advent of widespread hybrid Bohemian knotweed, regional differences, and the potential for biocontrol via the psyllid Aphalara itadori Shinji." Invasive Plant Science and Management 9.1 (2016): 60-70.
- Del Tredici, Peter. "The introduction of Japanese knotweed, Reynoutria japonica, into North America." The Journal of the Torrey Botanical Society 144.4 (2017): 406-416.
- EDDMapS. 2021. Early Detection & Distribution Mapping System. The University of Georgia - Center for Invasive Species and Ecosystem Health. Available online at http://www.eddmaps.org/; last accessed February 2, 2021.
- Fennell, Mark, Max Wade, and Karen L. Bacon. "Japanese knotweed (Fallopia japonica): an analysis of capacity to cause structural damage (compared to other plants) and typical rhizome extension." PeerJ 6 (2018): e5246.
- Invasive of the Week: Japanese Knotweed. Copyright © 2021 Matthaei Botanical Gardens & Nichols Arboretum. https://mbgna.umich.edu/invasive-of-the-week-japanese-knotweed/
- Japanese Knotweed – A brief history in time. https://www.eddmaps.org/species/subject.cfm?sub=19655
- Patocka, Jiri, Zdenka Navratilova, and M. A. R. I. B. E. L. Ovando. "Biologically active compounds of knotweed (Reynoutria spp.)." Mil. Med. Sci. Lett 86.1 (2017): 17-31.
- Peterson, Lee. A field guide to edible wild plants of eastern and central North America. No. 23. Houghton Mifflin Harcourt, 1977.
- Townsend, Ann. "Japanese knotweed: a reputation lost." Arnoldia 57.3 (1997): 13-19.
- Winston, R.L., C.B. Randall, B. Blossey, P.W. Tipping, E.C. Lake, and J. Hough-Goldstein. 2017. Field Guide for the Biological Control of Weeds in Eastern North America. USDA Forest Service, Forest Health Technology Enterprise Team, Morgantown, West Virginia. FHTET-2016-04.

Jimsonweed

- Cannas, A. "Department of Animal Science-Plants Poisonous to Livestock [Internet]. Cornel University. 2015 [cited 2018 Apr 14]."
- Carr, Anna. Rodale's illustrated encyclopedia of herbs. Rodale, 1998.
- Cheeke P. R. 1998. Natural Toxicants in Feeds, Forages, and Poisonous Plants. p. 382-383. 2nd. Ed. Interstate Pub. Inc. Danville, Illinois.

- EDDMapS. 2021. Early Detection & Distribution Mapping System. The University of Georgia - Center for Invasive Species and Ecosystem Health. Available online at http://www.eddmaps.org/; last accessed February 3, 2021.
- Hardin J. W. 1966. Stock-Poisoning Plants of North Carolina. p. 98-99. Bulletin No. 414. Agricultural Experiment Stat. North Carolina State Univ. Raleigh, NC.
- Jones, Hugh. The Present State of Virginia. No. 5. Reprinted for Joseph Sabin, 1865.
- Meyer, Joseph Ernest. The Herbalist and Herb Doctor. Indiana Herb Gardens, 1918.
- Muenscher W. C. 1946. Weeds. p. 406-408. The Macmillan Co. New York, New York.
- National Drug Intelligence Center 8201 Greensboro Drive, Suite 1001 McLean, VA 22102-3840 http://www.usdoj.gov/ndic/pubs/579/#Addresses
- Reader's Digest Association. Magic and medicine of plants. Reader's Digest, 1986.
- Sakonnakhon, S. Promsakha Na, et al. "Weeds–friend or foe? The role of weed composition on stover nutrient recycling efficiency." Field Crops Research 97.2-3 (2006): 238-247.

Kudzu

- Alderman, Derek H. "Channing Cope and the making of a miracle vine." Geographical Review 94.2 (2004): 157-177.
- Allen, A. B. 1843. Commercial nursery and garden of Messers. Parsons & Co. American Agriculturist 2(5): 130–132.
- Blaustein, Richard J. "Kudzu's invasion into Southern United States life and culture." The Great Reshuffling: Human dimensions of invasive species (2001): 55-62.
- https://conservingcarolina.org/get-rid-of-kudzu/
- Del Tredici, Peter. "The introduction of Japanese plants into North America." The Botanical Review 83.3 (2017): 215-252.
- Del Tredici, Peter. "From temple to terrace: The remarkable journey of the oldest bonsai in America." Arnoldia 64.2/3 (2006): 2-30.
- EDDMapS. 2021. Early Detection & Distribution Mapping System. The University of Georgia - Center for Invasive Species and Ecosystem Health. Available online at http://www.eddmaps.org/; last accessed February 3, 2021.
- Everest, J.W., J.H. Miller, D.M. Ball, and M. Patterson. 1999. Kudzu in Alabama: History, Uses, and Control (PDF | 1.46 MB)(link is external) Alabama Cooperative Extension System. ANR-65.
- Forseth, Irwin N., and Anne F. Innis. "Kudzu (Pueraria montana): history, physiology, and ecology combine to make a major ecosystem threat." Critical reviews in plant sciences 23.5 (2004): 401-413.
- Gant, Batsaki. "The Secret Life of Kudzu: A plant scorned as an invasive species offers a wealth of practical uses and cultural significance." Scientific American, 2019. https://blogs.scientificamerican.com/observations/the-secret-life-of-kudzu/
- https://www.historynet.com/kudzu-japans-wonder-vine.htm
- Hobbs, Richard J., ed. Invasive species in a changing world. Island Press, 2000.
- Miller, James H. 2003. Nonnative invasive plants of southern forests: a field guide for identification and

control. Gen. Tech. Rep. SRS–62. Asheville, NC: U.S. Department of Agriculture, Forest Service, Southern Research Station. 93 p.
- Miller, James H. "Kudzu eradication and management." In: Hoots:, Diane; Baldwin, Juanitta, comps., eds. Kudzu the vine to love or hate. Kodak, TN: Suntop Press: 137-149. (1996).
- Winberry, John J., and David M. Jones. "Rise and decline of the" miracle vine": kudzu in the southern landscape." Southeastern Geographer 13.2 (1973): 61-70.

Mullein

- Carr, Anna. Rodale's illustrated encyclopedia of herbs. Rodale, 1998.
- Darlington, William. American weeds and useful plants: being a second and illustrated edition of Agricultural botany. AO Moore, 1859.
- DeBray, L. The Wild Garden. 1978. Mayflower Books, Inc., New Yak.
- Durant, M. 1976. Who Named the Daisy? Who Named the Rose? Dodd, Mead & Company, New York.
- EDDMapS. 2021. Early Detection & Distribution Mapping System. The University of Georgia - Center for Invasive Species and Ecosystem Health. Available online at http://www.eddmaps.org/; last accessed February 3, 2021.
- Georgé, Stéphane, et al. "Changes in the contents of carotenoids, phenolic compounds and vitamin C during technical processing and lyophilisation of red and yellow tomatoes." Food Chemistry 124.4 (2011): 1603-1611.
- Gross, Katherine L. "Colonization by Verbascum thapsus (mullein) of an old-field in Michigan: experiments on the effects of vegetation." The Journal of Ecology (1980): 919-927.
- Lavin, Matt. "Invasive species compendium." (2016).
- Ody, Penelope. The complete medicinal herbal: a practical guide to the healing properties of herbs. Simon and Schuster, 2017.
- Reader's Digest Association. Magic and medicine of plants. Reader's Digest, 1986.
- Riaz, Muhammad, Muhammad Zia-Ul-Haq, and Hawa ZE Jaafar. "Common mullein, pharmacological and chemical aspects." Revista brasileira de farmacognosia 23.6 (2013): 948-959.
- Sakonnakhon, S. Promsakha Na, et al. "Weeds–friend or foe? The role of weed composition on stover nutrient recycling efficiency." Field Crops Research 97.2-3 (2006): 238-247.
- Silverman, Maida. A city herbal. Random House, 1977.
- Stevens, O. A. "Weights of seeds and numbers per plant." Weeds 5.1 (1957): 46-55.
- Turker, Arzu Ucar, and Ekrem Gurel. "Common mullein (Verbascum thapsus L.): recent advances in research." Phytotherapy Research: An International Journal Devoted to Pharmacological and Toxicological Evaluation of Natural Product Derivatives 19.9 (2005): 733-739.

Pigweed

- American Indian Health and Diet Project. https://aihd.ku.edu/foods/Pigweed.html
- https://www.ancientgrains.com/amaranth-history-and-origin/
- Arreguez, Guillermo A., Jorge G. Martínez, and Graciela Ponessa. " in an Archaeological Site from the Initial Mid-Holocene in the Southern Argentinian Puna Hybridus L. ssp.Amaranthus Hybridus." Quaternary International 307 (2013): 81–85, doi:10.1016/j.quaint.2013.02.035

- Banks, S. "Ministry of Agriculture, Food and Rural Affairs: Factsheet Pigweeds (Redroot, Green and Smooth)." Queen's Printer For Ontario 2016
- Carmack, Robert M., Janine L. Gasco, and Gary H. Gossen, eds. The legacy of Mesoamerica: history and culture of a Native American civilization. Routledge, 2016.
- Daehler, Curtis C. "The taxonomic distribution of invasive angiosperm plants: ecological insights and comparison to agricultural weeds." Biological Conservation 84.2 (1998): 167-180.
- EDDMapS. 2021. Early Detection & Distribution Mapping System. The University of Georgia - Center for Invasive Species and Ecosystem Health. Available online at http://www.eddmaps.org/; last accessed February 4, 2021.
- Foods of the Américas: Amaranth, the Outlaw Grain. María Elena Gaitán, Nov 18, 2017. https://cholaconcello.medium.com/foods-of-the-americas-amaranth-the-outlaw-grain-946957d9e51
- Graham, P. L., J. L. Steiner, and A. F. Wiese. "Light Absorption and Competition in Mixed Sorghum-Pigweed Communities." Agronomy journal 80.3 (1988): 415-418.
- Karttunen, Frances. An Analytical Dictionary of Nahuatl, University of Oklahoma Press,1992. 82.
- https://www.wildedible.com/pigweed-amaranth
- Minnesota Extension: Palmer Amaranth - a new weed to watch out for. Minnesota Extension, https://blog-crop-news.extension.umn.edu/2016/08/palmer-amaranth-new-weed-threat-to.html
- Mitich, Larry W. "Redroot pigweed (Amaranthus retroflexus)." Weed technology 11.1 (1997): 199-202.
- Sangorache, an Amaranth Used Ceremonially in Ecuador. American Anthropologist 66(1964): 136-140.
- Schonbeck, Mark. Weed Profile: Pigweeds (Amaranthus spp.). National Organic Program December 2019. https://eorganic.org/node/5120
- Tisler, A. M. 1990. Feeding in the pigweed flea beetle, Disonycha glabrata Fab. (Coleoptera: Chrysomelidae), on Amaranthus retroflexus. Virginia Journal of Science 41: 243–245.
- Vazin, F. "The effects of pigweed redroot (Amaranthus retoflexus) weed competition and its economic thresholds in corn (Zea mays)." Planta daninha 30.3 (2012): 477-485.

Plantain

- Ackley, Bruce A. Interactions of purple deadnettle, Lamium purpureum, soybean cyst nematode, Heterodera glycines and italian ryegrass, Lolium multIflorum. Diss. The Ohio State University, 2013.
- Cardona, F., Andrés-Lacueva, C., Tulipani, S., Tinahones, F., & Queipo-Orutño, M.I. (2013). Benefits of polyphenols on gut microbiota and implications in human health. The Journal of Nutritional Biochemistry, 24(8), 1415-1422. https://doi.org/10.1016/j.jnutbio.2013.05.001
- Creech, J. Earl, et al. "First report of soybean cyst nematode reproduction on purple deadnettle under field conditions." Crop Management 4.1. Purdue University, 2005.
- Culpeper, Nicholas. "The Complete Herbal (1653)." Reprinted in Birmingham, England by the Kynoch Press for the publishers Imperial Chemical (Pharmaceutical) Ltd (1953).

- Davis, D., Epp, M., & Riordan, H. 2004. Changes in USDA food composition data for 43 garden plants, 1950-1999. Journal of the American College of Nutrition, 23(6), 669-682. https://doi.org/10.1080/07315724.2004.10719409
- Deer, T.S. (2016). Allergy sufferers get ahead with purple dead-nettle. Retrieved from https://wisdomoftheplantdevas.com/tag/medicinal-plants-2/
- EDDMapS. 2021. Early Detection & Distribution Mapping System. The University of Georgia - Center for Invasive Species and Ecosystem Health. Available online at http://www.eddmaps.org/; last accessed February 5, 2021.
- Fraga, C.G., Galleano, M., Verstraeten, S.V., & Oteiza, P.I. (2010). Basic biochemical mechanisms behind the health benefits of polyphenols. Molecular Aspects of Medicine, 31(6), 435-445. https://doi.org/10.1016/j.mam.2010.09.006
- Grieve, M. (1971). A modern herbal Vols. 1-2. New York, NY: Dover Publications. (Original work published 1931)
- Holm, L., J. Doll, E. Holm, J. Pancho, and J. Herberger. 1997. World Weeds: Natural Histories and Distribution. New York: John Wiley and Sons. 1,129 p.
- Lavin, Matt. "Invasive species compendium." (2016).
- Magee, D.W., & Ahles, H.E. 1999. Flora of the Northeast. Amherst, MA: University of Massachusetts Press.
- Milburn, M. 2004. Indigenous nutrition: Using traditional food knowledge to solve contemporary health problems. American Indian Quarterly, 28(3-4), 411-434. https://doi.org/10.1353/aiq.2004.0104
- Purple Deadnettle Control: Getting Rid Of Deadnettle Weeds https://www.gardeningknowhow.com/plant-problems/weeds/purple-deadnettle-control.htm
- Reader's Digest Association. Magic and medicine of plants. Readers Digest, 1986.
- Steckel, Larry. Purple Deadnettle and Henbit. Tennessee Extension
- Webb, Jared S. The influence of winter annual weed control on soybean cyst nematode and summer annual weed growth and management. Southern Illinois University at Carbondale, 2007.
- Webster, T. M. 2000. Weed Survey - Southern States; Grass Crops Subsection. Proc. S. Weed Sci. Soc. 53:247-274.
- Wilson, B. J., N.C.B. Peters, K. J. Wright and H. A. Atkins. 1988. The influence of crop competition on seed production of Lamium purpureum, Vioia arvensis and Papaver rhose in winter wheat. Asp. Appl. Biol. 45:71-80

Puncturevine

- Adom, Muhammad Bahrain, et al. "Chemical constituents and medical benefits of Plantago major." Biomedicine & Pharmacotherapy 96 (2017): 348-360.

- Angier, Bradford. Field guide to edible wild plants. Stackpole Books, 2008.
- https://awkwardbotany.com/tag/plantain
- Blair, K. (2014). The wild wisdom of weeds: 13 essential plants for human survival. White River Junction, Vermont: Chelsea Green Publishing.
- Carr, Anna. Rodale's illustrated encyclopedia of herbs. Rodale, 1998.
- Dyer, T.S. Thistleton. The Folklore of Plants. 1889.
- EDDMapS. 2021. Early Detection & Distribution Mapping System. The University of Georgia - Center for Invasive Species and Ecosystem Health. Available online at http://www.eddmaps.org/; last accessed February 4, 2021.
- Gardening Know How: History Of Plantain https://blog.gardeningknowhow.com/tbt/history-of-plantain-tbt/
- Guil-Guerrero, J. L. "Nutritional composition of Plantago species (P. major L., P. lanceolata L., and P. media L.)." Ecology of food and nutrition 40.5 (2001): 481-495.
- Jarić, Snežana, et al. "Traditional Wound-Healing Plants Used in the Balkan Region (Southeast Europe)." Journal of Ethnopharmacology, vol. 211, Jan. 2018, pp. 311–328.
- Kinney, Shirley. "Anglo-Saxon Medicine: Cures or Catastrophe?." Primary Source 2: 37-42.
- Meyer, Joseph Ernest. The Herbalist and Herb Doctor. Indiana Herb Gardens, 1918.
- Moss, Kay. Southern Folk Medicine, 1750-1820. University of South Carolina Press, 1999.
- Nafici, Saara. Weed of the Month: Broadleaf Plantain. Brooklyn Botanic Garden 2014. https://www.bbg.org/news/weed_of_the_month_broadleaf_plantain
- https://www.northeastipm.org/index.cfm/schools/pests/broadleaf-plantain/
- Ody, Penelope. The complete medicinal herbal: a practical guide to the healing properties of herbs. Simon and Schuster, 2017.
- Olsen, Karin. "The Lacnunga and Its Sources: The Nine Herbs Charm and Wið Færstice Reconsidered." (2007).
- Reader's Digest Association. Magic and medicine of plants. Reader's Digest, 1986.
- Roth, Sally. Weeds, Friend Or Foe?. Reader's Digest, 2002.
- Shakespeare, William. Romeo and juliet. Vol. 1. Classic Books Company, 2000.
- Samuelsen, A. B. The traditional uses, chemical constituents and biological activities of Plantago major L. A review. Journal of Ethnopharmacology, 1 jan. 2000. v. 71, p. 1–21.
- Sieling, Peter. "Chapter Five: Appalachian Folk Remedies." Folk Medicine, Jan. 2003, p. 20.
- Silverman, Maida. A city herbal. Distributed by Random House, 1977.
- Watts, Donald. Dictionary Of Plant Lore. Amsterdam: Academic Press, 2007. eBook Collection (EBSCOhost).

Purple Dead Nettle

- Boydston, R. A. 1990. Time and emergence and seed production of longspine sandbur (Cenchrus longispinus) and puncturevine (Tribulus terrestris). Weed Sci. 38:16-21.
- Donaldson, S. and D. Rafferty. 2003. Identification and management of puncturevine (Tribulus terrestris L.). Univ. of Nevada Coop. Ext. Serv. Circ. Fact Sheet FS-03-34. Reno, NV.
- EDDMapS. 2021. Early Detection & Distribution Mapping System. The University of Georgia - Center for

Invasive Species and Ecosystem Health. Available online at http://www.eddmaps.org/; last accessed February 4, 2021.
- Goathead.com – Commercial Source for Puncturevine Weevils. Information about the Plant and the Weevils. http://www.goatheads.com/
- Great Plains Flora Association. 1986. Flora of the Great Plains. Lawrence, KS: Univ. Press of Kansas.1392 pp.
- Johnson, E. 1932. The puncturevine in California. Univ. of Calif. Agric. Expt. Sta. Bull. 528: 42 pp.
- List, Uses, and Host Plant. "Tribulus terrestris (puncture vine)."
- https://www.nwcb.wa.gov/weeds/puncturevine
- Parker, Robert. Puncturevine. Pacific Northwest Extension Publications 1983.
- Parsons WT and Cuthbertson EG (1992). Noxious Weeds of Australia. Inkata Press, Melbourne, Australia.
- Puncturevine bur and flower photographs courtesy of Virginia Tech Weed Guide.
- Puncturevine Images – U-GA Bugwood Image Gallery.
- Puncturevine photograph courtesy of University of Nevada Cooperative Extension Service.
- Puncturevine Profile – CDFA Encycloweedia. http://www.cdfa.ca.gov/phpps/ipc/weedinfo/tribulus-terrestris.htm
- Puncturevine Profile – U-CA IPM Online. http://www.ipm.ucdavis.edu/PMG/PESTNOTES/pn74128.html
- Puncturevine Profile – USDA Plants Database. http://plants.usda.gov/java/profile?symbol=trte
- Squires, V.R. 1979. The biology of Australian weeds. 1. Tribulus terrestris L. J. of the Australian Inst. of Agric. Sci. 179: 75-82.
- Westbrooks, Randy. EDRR Fact Sheet. U.S. National Early Detection and Rapid Response System for Invasive Plants
- Whitson, T. D., editor. 2000. Weeds of the West 9th Ed. Western Society of Weed Science, Newark, CA 94560. 630pp.

Purslane

- Balyan, R. S., and V. M. Bhan. 1986. "Emergence, Growth, and Reproduction of Horse Purslane (Trianthema portulacastrum) as Influenced by Environmental Conditions." Weed Science 34: 516–9.
- Chapman, Jefferson, Robert B. Stewart, and Richard A. Yarnell. "Archaeological evidence for pre-Columbian introduction of Portulaca oleracea and Mollugo verticillata into eastern North America." Econ. Bot 28.4 (1974): 411-412.
- EDDMapS. 2021. Early Detection & Distribution Mapping System. The University of Georgia - Center for Invasive Species and Ecosystem Health. Available online at http://www.eddmaps.org/; last accessed February 5, 2021.
- https://www.illinoiswildflowers.info/weeds/plants/purslane.htm
- Lavin, Matt. "Invasive species compendium." (2016).
- Montgomery, E H. 1964. Weeds of Canada and the Northern United States. Toronto: Ryerson Press. 266 p.
- Okafor, Izuchukwu Azuka, Mary B. Ayalokunrin, and Lovina Abu Orachu. "A review on Portulaca oleracea (purslane) plant–its nature and biomedical benefits." International journal of Biomedical research 5.2 (2014): 75-

80.
- Southern Weed Science Society (SWSS). 1993. Weed Identification Guide. Champaign: Southern Weed Science Society.
- "Trianthema portulacastrum L. - AIZOACEAE – Dicotyledon." Accessed December 19, 2019.http://idao.cirad.fr/content/oscar/especes/t/trtpo/trtpo.html
- Turbelin, Anna J., Bruce D. Malamud, and Robert A. Francis. "Mapping the global state of invasive alien species: patterns of invasion and policy responses." Global Ecology and Biogeography 26.1 (2017): 78-92.
- Uddin, Md, et al. "Purslane weed (Portulaca oleracea): a prospective plant source of nutrition, omega-3 fatty acid, and antioxidant attributes." The Scientific World Journal 2014 (2014).
- Watson, P. J. 1969. The prehistory of Salts Cave, Kentucky. Reports of Investigations, No. 16. Illinois State Museum, Springfield.
- https://wssa.net/wp-content/themes/WSSA/WorldOfWeeds/purslane.html

Siberian Elm

- Christensen, Earl M. "The recent naturalization of Siberian elm (Ulmus pumila L.) in Utah." The Great Basin Naturalist (1964): 103-106.
- Ding, Jianqing, et al. "Biological control of invasive plants through collaboration between China and the United States of America: a perspective." Biological invasions 8.7 (2006): 1439-1450.
- Hirsch, Heidi. "Investigations on the invasion success of Ulmus pumila-a multi-continental survey." (2013).
- Leopold, Donald J. "Chinese and Siberian elms." Journal of Arboriculture 6.7 (1980): 175-179.
- McIlvain, E. H., and C. G. Armstrong. "Siberian Elm—A Tough New Invader of Grasslands." Weeds 13.3 (1965): 278-279.
- Moss, Rebecca. Rethinking the dreaded Siberian elm. The Santa Fe New Mexican, May 13, 2017
- Rehder, A. 1923. Enumeration of the ligneous plants of northern China. Jour. Arnold Arb. 4:167-168.
- Sander, D. H. "Soil properties and Siberian elm tree growth in Nebraska windbreaks." Soil Science 112.5 (1971): 357-363.
- SIBERIAN ELM (Ulmus pumila). North Dakota State Extension Fact Sheet. https://www.nd.gov/ndda/sites/default/files/legacy/resource/SIBERIANELM.pdf
- The Texas Invasive Database, https://www.texasinvasives.org/plant_database
- https://plants.usda.gov/factsheet/pdf/fs_ulpu.pdf
- USDA Field Guide for Managing Siberian Elm in the Southwest, https://www.fs.usda.gov/Internet/FSE_DOCUMENTS/stelprdb5410128.pdf

Sow Thistle

- https://www.canr.msu.edu/weeds/extension/perennial-sowthistle
- Chauhan, Bhagirath S., Gurjeet Gill, and Christopher Preston. "Factors affecting seed germination of

annual sowthistle (Sonchus oleraceus) in southern Australia." Weed Science 54.5 (2006): 854-860.
- EDDMapS. 2021. Early Detection & Distribution Mapping System. The University of Georgia - Center for Invasive Species and Ecosystem Health. Available online at http://www.eddmaps.org/; last accessed February 6, 2021.
- McClay, A. S., and D. P. Peschken. "80 Sonchus arvensis L., Perennial Sow-thistle (Asteraceae)." Biological Control Programmes in Canada, 1981-2000 (2001): 416.
- https://www.nwcb.wa.gov/weeds/perennial-sowthistle
- https://www.nwcb.wa.gov/images/weeds/PERENNIAL-SOWTHISTLE-BROCHURE_Lincoln.pdf
- Roth, Sally. Weeds, Friend Or Foe?. Reader's Digest, 2002.
- Silverman, Maida. city herbal. Distributed by Random House, 1977.
- USDA Forest Service Perennial Sowthistle Leaflet, https://www.fs.usda.gov/detail/r10/communityforests/?cid=fseprd569328

Tree of Heaven

- https://www.botanical.com/botanical/mgmh/t/trehea28.html
- Tree of Heaven: An Exotic Invasive Plant Fact Sheet. Ecological Landscape Alliance, https://www.ecolandscaping.org/05/landscape-challenges/invasive-plants/tree-of heaven-an-exotic-invasive-plant-fact-sheet/
- EDDMapS. 2021. Early Detection & Distribution Mapping System. The University of Georgia - Center for Invasive Species and Ecosystem Health. Available online at http://www.eddmaps.org/; last accessed February 6, 2021.
- Kummer, Frank. Why the tree of heaven spreads so devilishly and harms Pa. forests. The Philadelphia Enquirer, August 2, 2017.
- Mori, Scott A. Tree of Heaven: An Immigrant Thriving in New York and Beyond. New York Botanic Gardens 2015
- Nature Conservancy. Tree of Heaven. July 06 2020. https://www.nature.org/en-us/about-us/where-we-work/united-states/indiana/stories-in-indiana/journey-with-nature—tree of heaven
- Penn State Extension Invasive Plant Sheet: Tree of Heaven. https://extension.psu.edu/tree of heaven
- https://extension.umd.edu/sites/extension.umd.edu/files/_docs/programs/woodland-steward/DNR_TreeOfHeaven.pdf
- Virginia Department of Forestry: Control and Utilization of tree of heaven: A Guide for Virginia Landowners. http://www.dof.virginia.gov/infopubs/Control-and-Utilization-of-tree of heaven-2019-03_pub.pdf

Tumbleweed

- Tumbleweed: Russian Thistle. Desert USA. https://www.desertusa.com/flowers/tumbleweed.html
- DiTomaso, J. M., and E. A. Healy. 2007.Weeds of California and other Western States. Oakland: Univ.

Calif. Agric. Nat. Res. Publ. 3488.Elias, Emile, et al. "Climate change, agriculture and water resources in the Southwestern United States." Journal of Contemporary Water Research & Education 158.1 (2016): 46-61.
- Hickman, J. C., ed. 1993. Jepson Manual: Higher Plants of California. Berkeley: Univ. Calif. Press.
- Muller, Nora. Chasing Down the History of the Tumbleweed. Garden Collage Magazine January 2016. https://gardencollage.com/wander/gardens-parks/chasing-history-tumbleweed/
- Ndée Bikíyaa – The People's Farm. https://www.firstnations.org/videos/ndee-bikiyaa-the-peoples-farm/
- Nuwer, Rachel. America's Tumbleweeds Are Actually Russian Invaders. Smithsonian Magazine August 2014. https://www.smithsonianmag.com/smart-news/americas-tumbleweeds-are-actually-russian-invaders-180952360/
- Traditional Ecological Knowledge: Southwest. https://www.nps.gov/subjects/tek/southwest.htm
- Rhodes, W. A., E. F. Frolich, and A. Walice. 1967. Russian thistle seeds.Calif. Agric. 21(4):2.
- RUSSIAN THISTLE: Integrated Pest Management in the Landscape. UC Statewide IPM Program, University of California
- Smith, L., R. Sobhian, and M. Cristofaro. 2006. Prospects for biological control of Russian thistle (tumbleweed). California Invasive Plant Council Symposium Nov. 1, 2006. 41:120-133. Available online from the United States Department of Agriculture-Agriculture Research Service. Accessed Dec 12, 2007.
- Whitson, T. D., L. C. Burrill, S. A. Dewey, D. W. Cudney, B. E. Nelson, R. D. Lee, and R. Parker. 2002. Weeds of the West. Jackson, WY: Univ. Wyoming and Western Society of Weed Science.
- Wills, Wirt H. "Early agriculture and sedentism in the American Southwest: Evidence and interpretations." Journal of World Prehistory 2.4 (1988): 445-488.
- Young, James A. "Tumbleweed." Scientific American 264.3 (1991): 82-87.

Two-Spotted Spurge

- Asgarpour, Rayhaneh, et al. "Germination of spotted spurge (Chamaesyce maculata) seeds in response to different environmental factors." Weed Science 63.2 (2015): 502-510.
- https://www.bellarmine.edu/faculty/drobinson/prostratespurge.asp
- EDDMapS. 2021. Early Detection & Distribution Mapping System. The University of Georgia - Center for Invasive Species and Ecosystem Health. Available online at http://www.eddmaps.org/; last accessed February 6, 2021.
- HISTORY OF EARLY AMERICAN LANDSCAPE DESIGN: A Project of the Center for Advanced Study in the Visual Arts, National Gallery of Art. https://heald.nga.gov/mediawiki/index.php/Nursery
- Montana field guides. http://fieldguide.mt.gov/speciesDetail.aspx?elcode=PDEUP0D1E0
- NAL Special Collections, United States Department of Agriculture. Prince Family Manuscript Collection. https://specialcollections.nal.usda.gov/guide-collections/prince-family-manuscript-collection
- Nebraska Extension: Community Environment; Prostrate Spurge. https://communityenvironment.unl.edu/prostrate-spurge
- Penn State University Extension, https://extension.psu.edu/lawn-and-turfgrass-weeds-spotted-spurge-chamaesyce-maculata-l
- https://plants.ces.ncsu.edu/plants/euphorbia-maculata/
- Roth, Sally. Weeds, Friend Or Foe?. Reader's Digest, 2002.

- University of Arkansas, Plant of the Week: Spurge, Prostrate. https://www.uaex.edu/yard-garden/resource-library/plant-week/spurge-prostrate-7-22-11.aspx
- University of California Agricultural and Natural Resources Statewide Integrated Pest Management Program-Spotted Spurge and Other Spurges. http://ipm.ucanr.edu/PMG/PESTNOTES/pn7445.html
- Uva, R.H., J.C. Neal, and J.M. DiThomaso. 1997. Weeds of the northeast. Cornell Univ. Press. 397 pp.
- Wildflowers, Minnesota. "A field guide to the flora of Minnesota." Vicia villosa (2020).

Wild Mustard

- Kew Science Plants of The World Online: Brassica oleracea L. http://powo.science.kew.org/taxon/urn:lsid:ipni.org:names:279435-1
- Ministry of Agriculture, Food and Rural Affairs Ontario. Wild Mustard Fact Sheet. http://www.omafra.gov.on.ca/english/crops/facts/03-043.htm
- http://www.mobileranger.com/blog/wild-mustard-creates-beautiful-views-and-tastes-good-too/
- National Park Service: Wild Mustard. https://www.nps.gov/prsf/learn/nature/wild-mustard.htm
- Ody, Penelope. The complete medicinal herbal: a practical guide to the healing properties of herbs. Simon and Schuster, 2017.
- Roth, Sally. Weeds, Friend Or Foe?. Reader's Digest, 2002.
- Siskin, Joshua. Gardening: There's a wild story behind wild mustard. Los Angeles Daily News August 28, 2017USDA, NRCS. 2021. The PLANTS Database (http://plants.usda.gov, 7 February 2021). National Plant Data Team, Greensboro, NC 27401-4901 USA.
- Wild mustard – Sinapis arvensis. Michigan State University, https://www.canr.msu.edu/resources/wild-mustard-sinapis-arvensis

Wild Parsnip

- Ontario Invasive Plant Council. Best Management Practice Technical Document for Land Managers. 2017.
- Zangerl, A. R., and May R. Berenbaum. "Effects of florivory on floral volatile emissions and pollination success in the wild parsnip." Arthropod-Plant Interactions 3.3 (2009): 181-191.
- Riha, Jacob. A review of the potential impacts and mitigation techniques of noxious weeds in the Lennox and Addington County. Diss. Queens University 2018. https://qspace.library.queensu.ca/bitstream/handle/1974/24209/ensc_501_riha.pdf?sequence=1&isAllowed=y
- Lavin, Matt. "Invasive species compendium." (2016).

- https://www.healthvermont.gov/health-environment/environmental-chemicals-pollutants/wild-poison-parsnip
- http://www.wrightswcd.org/docs/Trifold.pdf
- Jogesh, Tania, et al. "Patterns of genetic diversity in the globally invasive species Wild Parsnip (*Pastinaca sativa*)." Invasive Plant Science and Management 8.4 (2015): 415-429.
- Venema, Christine. "Parsnips: A vegetable from antiquity." Michigan State University Extension, September 25, 2015. https://www.canr.msu.edu/news/parsnips_a_vegetable_from_antiquity
- Carr, Anitra C., and Silvia Maggini. "Vitamin C and immune function." Nutrients 9.11 (2017): 1211.
- USDA Department of Agriculture - FoodData Central: "Parsnips, raw."
- Victoria State Government - Department of Health and Human Services: "Parsnips."
- Mayo Clinic - Healthy Lifestyle: Nutrition and healthy eating: "Dietary fiber: Essential for a healthy diet." https://www.mayoclinic.org/healthy-lifestyle/nutrition-and-healthy-eating/in-depth/fiber/art-20043983
- Oregon State University, Linus Pauling Institute: "Glycemic Index and Glycemic Load." https://lpi.oregonstate.edu/mic/food-beverages/glycemic-index-glycemic-load
- Harvard Medical School - Harvard Health Publishing: "The importance of potassium." https://www.health.harvard.edu/staying-healthy/the-importance-of-potassium
- New York State - Department of Environmental Conservation: "Wild Parsnip."
- https://www.oardc.ohio-state.edu/weedguide/single_weed.php?id=108
- USDA, NRCS. 2021. The PLANTS Database (http://plants.usda.gov, 7 February 2021). National Plant Data Team, Greensboro, NC 27401-4901 USA.
- EDDMapS. 2021. Early Detection & Distribution Mapping System. The University of Georgia - Center for Invasive Species and Ecosystem Health. Available online at http://www.eddmaps.org/; last accessed February 7, 2021.

Yellow Woodsorrel

- Blackburn, Joni. Weed of the Month: Yellow Wood-Sorrel. April 2, 2019. https://www.bbg.org/news/weed_of_the_month_yellow_wood_sorrel
- Meuninck, Jim. Basic Illustrated Edible Wild Plants and Useful Herbs. Rowman & Littlefield, 2018.
- Chandran, Rakesh.Yellow Woodsorrel. West Virginia University Extension. September 2020
- Reader's Digest Association. Magic and medicine of plants. Reader's Digest, 1986.
- Linex, Ricky. "Yellow Wood Sorrel." Ranch and Rural Living 99.8 (2018): 5-7.
- Saha, Debilina et al. Identifying and managing yellow woodsorrel (Oxalis stricta L.) in nurseries and greenhouses. Michigan State University Department of Horticulture, MSU Extension Bulletin E3440.
- Marble, S. Christopher, et al. "Early postemergence control of yellow woodsorrel (Oxalis stricta) with residual herbicides." Weed Technology 27.2 (2013): 347-351.
- Roth, Sally. Weeds, Friend Or Foe?. Reader's Digest, 2002.
- EDDMapS. 2021. Early Detection & Distribution Mapping System. The University of Georgia - Center for Invasive Species and Ecosystem Health. Available online at http://www.eddmaps.org/; last accessed February 7, 2021.

Index

262

forests, 78, 87, 116, 184, 247, 256
formulations
  chemical, 236
  solid, 236–37
Forseth, 249
Forty-Eighters, 56
founding father, 201
foxtail grass, 81, 83–84, 246
French Empire, 34
French king, 34
French traders, 33
fruits, 11, 21, 61, 64, 97, 102, 139, 147, 197, 205, 211

G

gallbladder problems, 66
gamma-linolenic acid, 164
Gant, Philip, 120
gardening practices, 230
garden pests, 230
garden plants, 56, 251
gardens, 9, 11, 28, 53, 55–59, 153, 155, 157, 200–201, 205, 207–8, 232, 235, 237
Gardner, 244
Gaspar, 207
genes, 213
genetic divergence, 56
genetic diversity, 247, 259
Georgia, 1, 18, 32, 119, 133, 241–45, 247–54, 256–57, 259
Georgia O'Keefe, 111
Gerard, 65
germinating, 112, 150, 163, 165, 209, 220
germination, 28, 151, 200
germination rate, 219
Giant Hogweed, 87, 93, 247
Gilbert, 245

glufosinate, 237
Glycemic Index, 259
Glycemic Load, 259
glyphosate, 9, 43, 50, 59, 93, 105, 122, 134, 136, 237, 245
  concentrated, 173
glyphosate-based herbicide, 185
glyphosate LD50, 59
glyphosate products, 93
goats, 48, 85, 110, 122, 129, 137, 149, 178, 185–86
Gobi Desert, 167
Gold Rush, 184
Graham, 251
grain, 41, 47, 83, 135, 142
Gram-negative bacteria, 128
Gramoxone, 237
Grazing, 78, 122, 129
Great Depression, 77, 119
Great Lakes states, 103
Great Plains, 241, 254
Greek philosopher, 65
Greeks, 3, 13, 127, 163
Greek word, 27, 219
greenhouses, 155, 219, 259
greens, 57, 77, 135–36
Greensboro, 258–59
groundcover, 129, 144, 220
  heat-tolerant, 165
growing season, 11, 42, 64, 81, 92–93, 143, 147, 161, 185, 209, 217
Guillermo, 250

H

habitat, 89, 200
  common, 149
habitat damage, 181
habitat protection, 248

mustard, 207–8

274

About The Author

Olivia Wylie is a professional landscaper who specializes in the restoration of neglected gardens. When the weather keeps her indoors, she enjoys researching and writing about the plants she loves and the ways they've shaped human thought. She lives in Colorado with a very patient husband and a rather impatient cat. Her works can be viewed at www.leafingoutgardening.com.

www.ingramcontent.com/pod-product-compliance
Lightning Source LLC
Chambersburg PA
CBHW040246100426
42811CB00011B/1167